3rd – 6th

Ready-to-Use

Multiplication Practice Puzzles

Written & Illustrated By
Marsha Sanger

Published By

"Leading The Way In Creative Educational Materials" ™

857 Lake Blvd. ❖ Redding, California 96003

TO THE TEACHER

These multiplication puzzles are meant to answer the question — *What do I do now?*. I assign a puzzle each Friday, after I give a weekly ten question math quiz. Students in my class enjoy completing these math puzzles and constanly ask for more...more...more. There are always some of these puzzles available to my students so they can earn extra credit.

The puzzles in this book are sequenced in alphabetical order. This was done so each puzzle and answer key could be found easily. However, I suggest that you categorize them according to month, as well as to topics being taught in your class. These activities were designed to give students something fun to do while at the same time practicing those *dreaded* multiplication facts.

The number possible for each problem is located in the upper-right corner of the puzzle's frame. A color key and answer key is provided for each puzzle. I allow students to correct their own completed puzzles. This not only saves me time, it gives students immediate feedback. It is important to note that there should never be any two shapes with the same color touching (except when noted in the school bus puzzle). If two shapes with the same color touch — there is a mistake.

(Note: If you want any of your students to use a multiplication grid, there is one on page 123.)

I hope your students enjoy doing these activities as much as mine. Have fun!

Marsha Sanger
Author

Copyright ©1996 Golden Educational Center
Revised 1999   All Rights Reserved – Printed in U.S.A.
Published By Golden Educational Center
857 Lake Blvd. ❖ Redding, California 96003

Notice
Reproduction of worksheets by the classroom teacher for use in the classroom and not for commercial sale is permissible.

No part of this publication may be reproduced, stored in a retrieval system, or transmitted, in any form or by any means, electronic, mechanical, recording or otherwise, without written permission of the publisher.

Reproduction of these materials for an entire school, or for a system or district is strictly prohibited.

ISBN 1-56500-040-4

Order of
Answer Keys (even numbers) & Puzzles (odd numbers)

Abraham Lincoln	2–3
Apple	4–5
Baseball, Bat & Cap	6–7
Bear	8–9
Bee	10–11
Bell	12–13
Bird	14–15
Bunny	16–17
Butterfly	18–19
Candles	20–21
Clipboard	22–23
Cosmos Flower	24–25
Cupid	26–27
Daffodil	28–29
Elephant	30–31
Fish	32–33
Flag	34–35
Frog	36–37
George Washington	38–39
Groundhog	40–41
Hamburger	42–43
Headdress (Native American)	44–45
Horse	46–47
Ice Cream	48–49
July 4th	50–51
Kite	52–53
Kitten	54–55
Ladybug	56–57
Leaf	58–59
Liberty Bell	60–61
Lily	62–63
Lion	64–65
Magnolia Flower	66–67
Moon	68–69
Mouse	70–71
Mushrooms	72–73
Numbers	74–75
Penguin	76–77
Pony	78–79
Poppy	80–81
Rabbit	82–83
Raccoon	84–85
School Bus	86–87
Scroll	88–89
Seal	90–91
Shamrock	92–93
Snowman	94–95
Squirrel	96–97
Star	98–99
Statue of Liberty	100–101
Stocking	102–103
Strawberry	104–105
Sunflower	106–107
Swans	108–109
Teddy Bear	110–111
Tepee	112–113
Tiger Cub	114–115
Tree	116–117
U.S.A.	118–119
Valentine	120–121

© Golden Educational Center

Multiplication Puzzles

Correction Key

Abraham Lincoln

1. Allow your students to correct their own work.
2. Make a transparency of this puzzle and instruct your students to place the transparency over their completed puzzle for a quick and easy check.

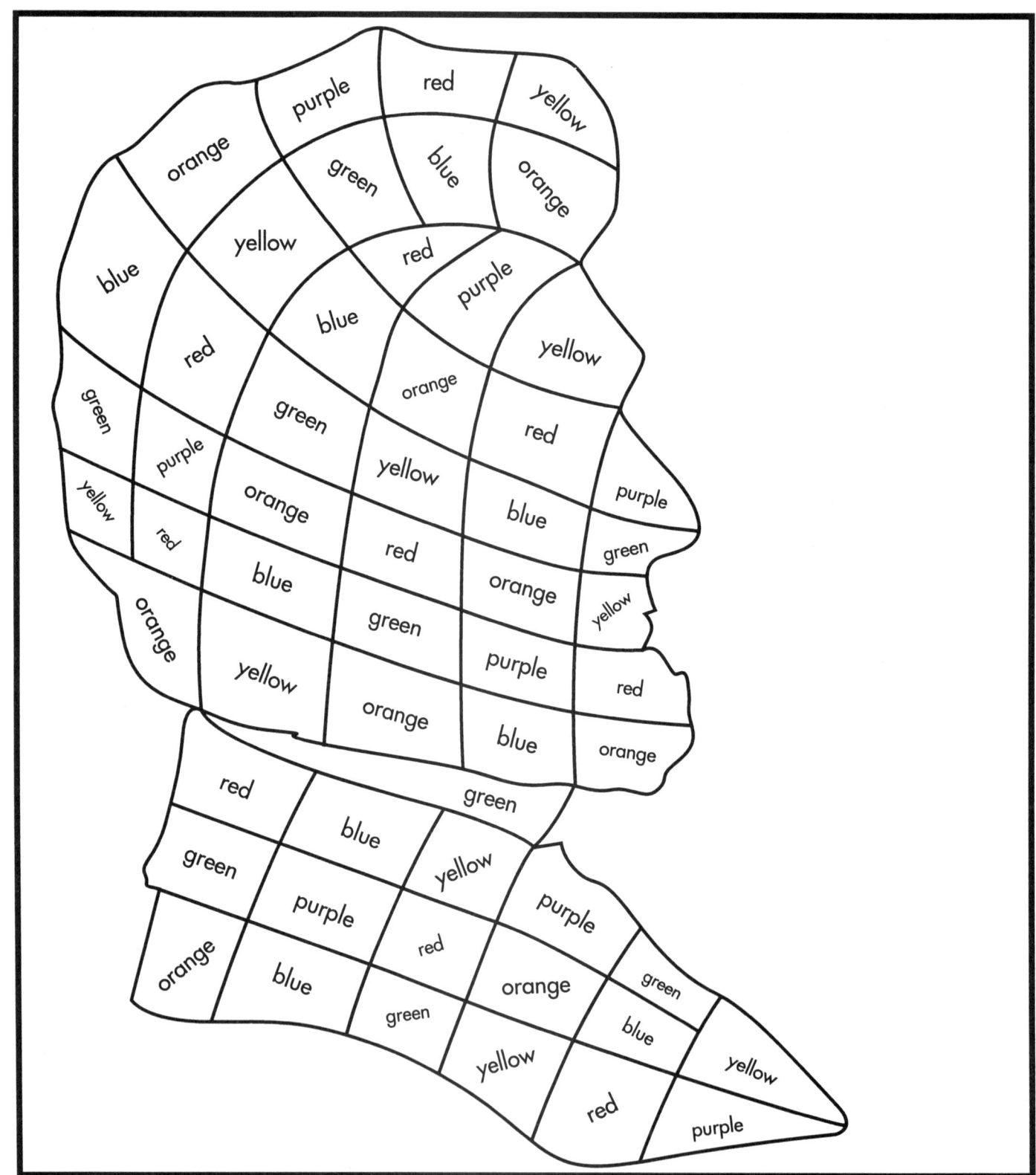

Multiplication Puzzles ✦ Activity 1

© Golden Educational Center

Abraham Lincoln

Name _____

Date _____

1. Complete the problems within the parentheses first. Then complete the subtraction problem using your answer.

2. Using your final answer and the color key, color your puzzle correctly.

56 Total Problems

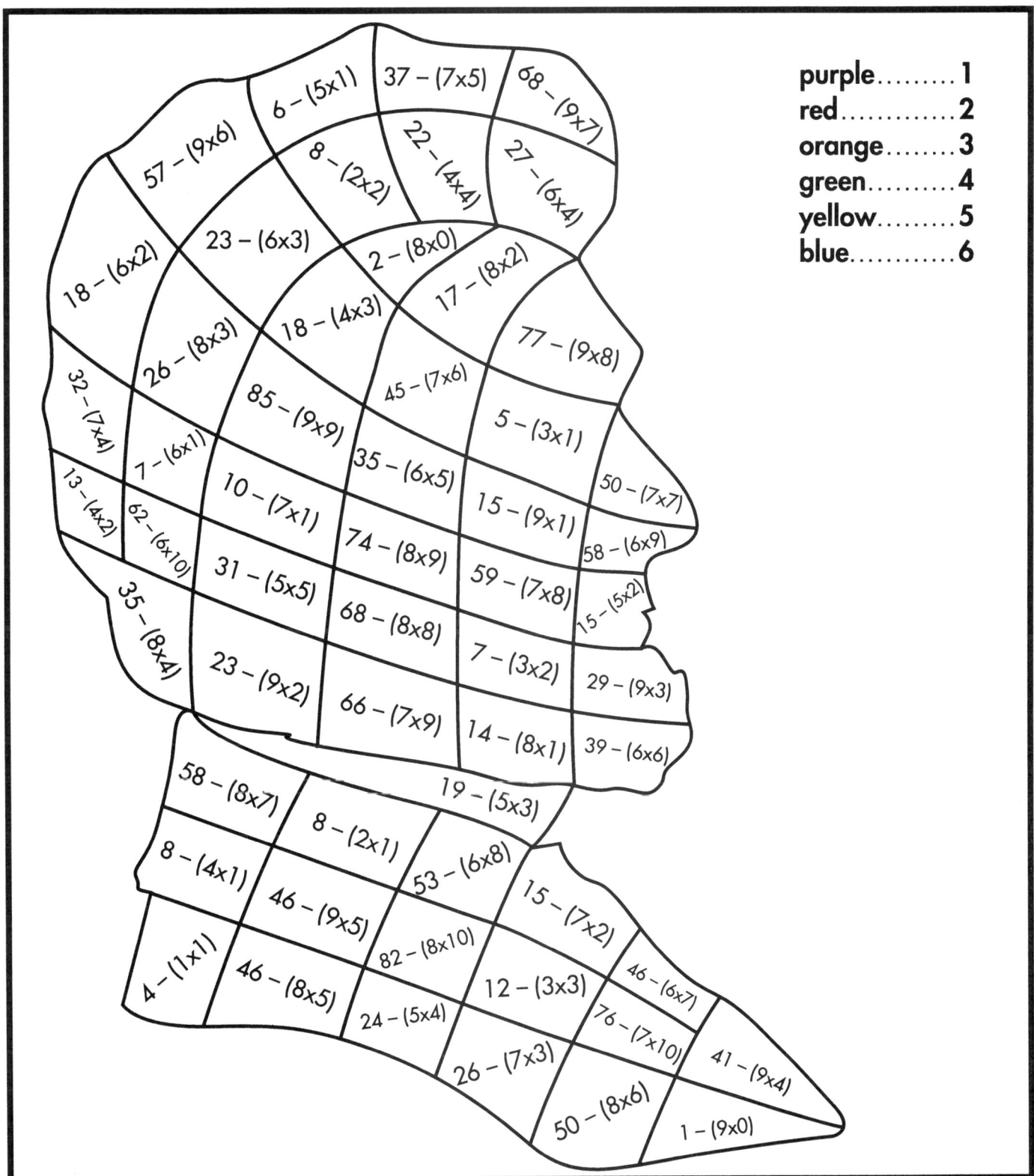

purple.........1
red.............2
orange........3
green..........4
yellow.........5
blue............6

© Golden Educational Center

Multiplication Puzzles ✦ **Activity 1**

Correction Key

Apple

1. Allow your students to correct their own work.
2. Make a transparency of this puzzle and instruct your students to place the transparency over their completed puzzle for a quick and easy check.

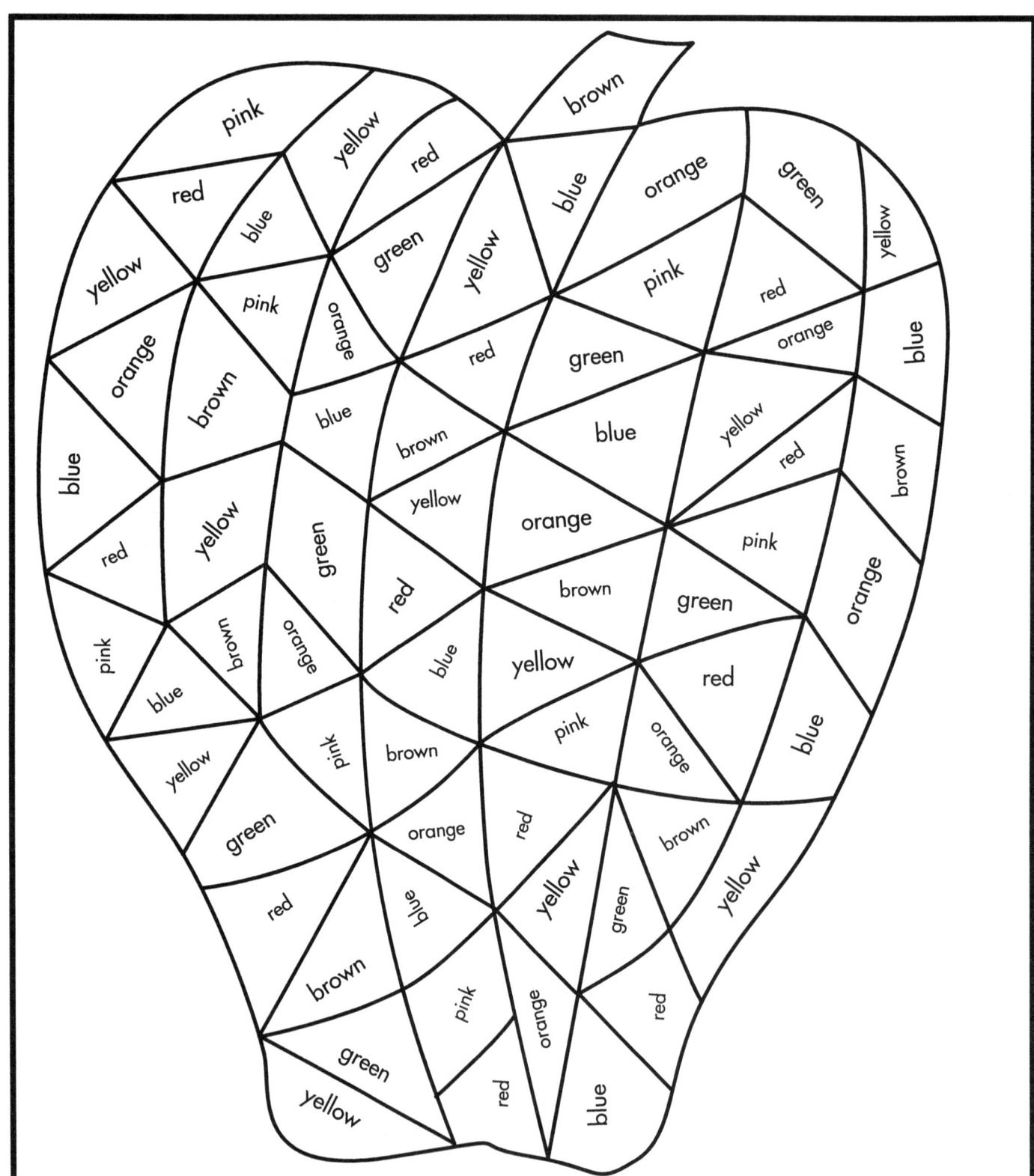

Multiplication Puzzles ✦ Activity 2

© Golden Educational Center

Apple

1. Complete the problems within the parentheses first. Then complete the subtraction problem using your answer.
2. Using your final answer and the color key, color your puzzle correctly.

70 Total Problems

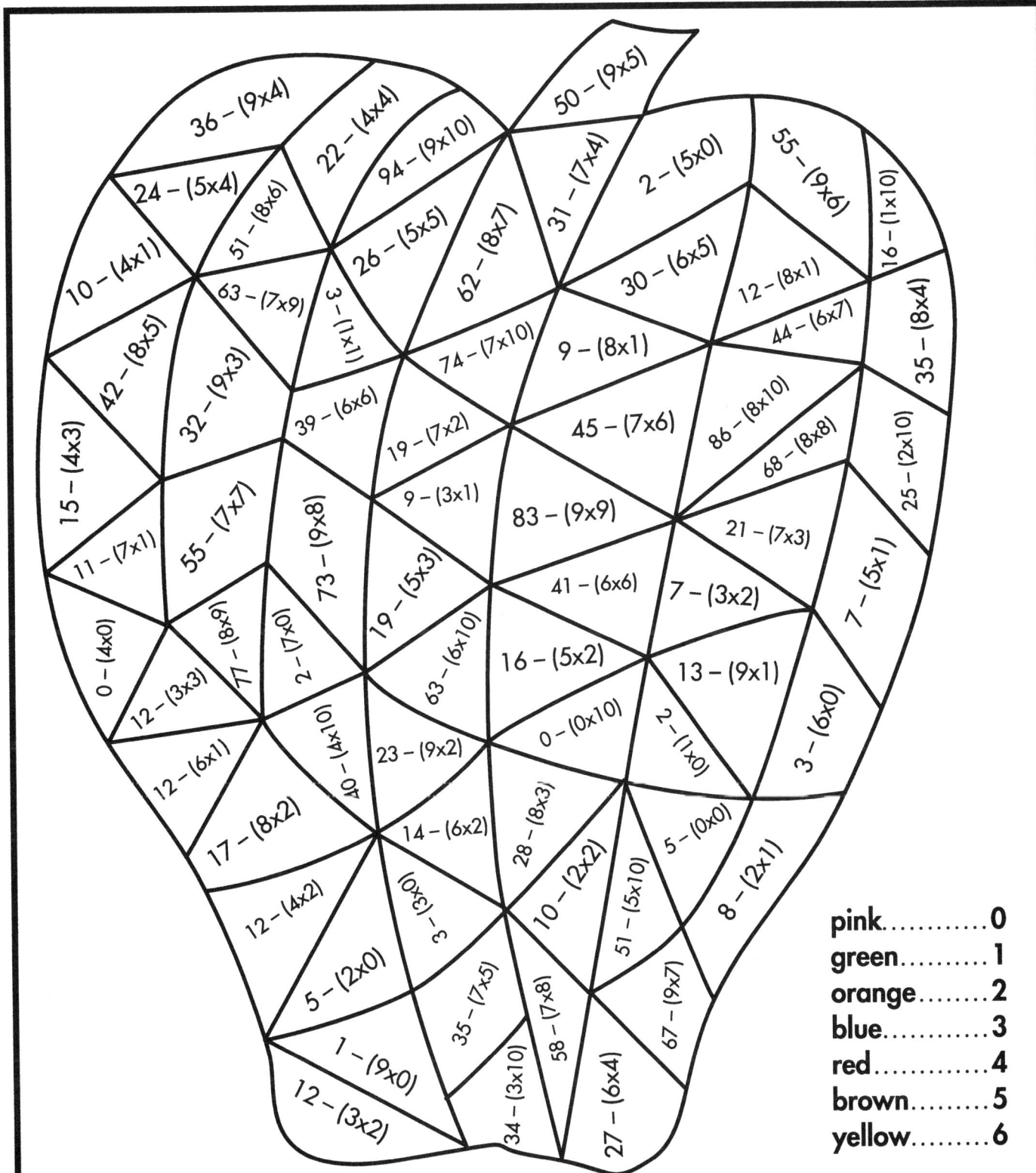

pink..........0
green.........1
orange........2
blue..........3
red...........4
brown.........5
yellow........6

Correction Key

Baseball

1. Allow your students to correct their own work.
2. Make a transparency of this puzzle and instruct your students to place the transparency over their completed puzzle for a quick and easy check.

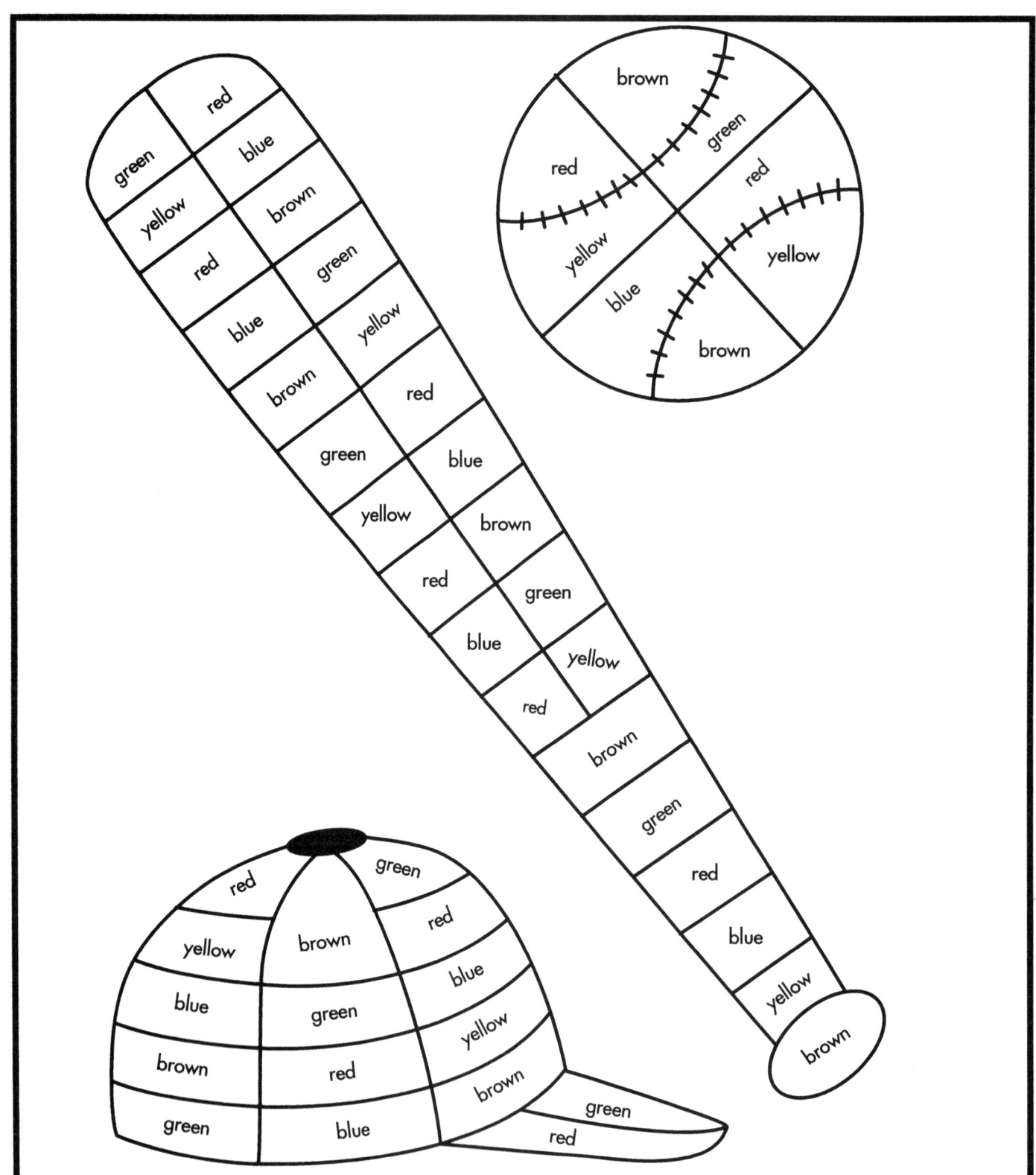

Multiplication Puzzles ✦ Activity 3

Baseball

Name _____

Date _____

1. Complete the problems within the parentheses first. Then complete the addition or subtraction problem using your answer.
2. Using your final answer and the color key, color your puzzle correctly.

50 Total Problems

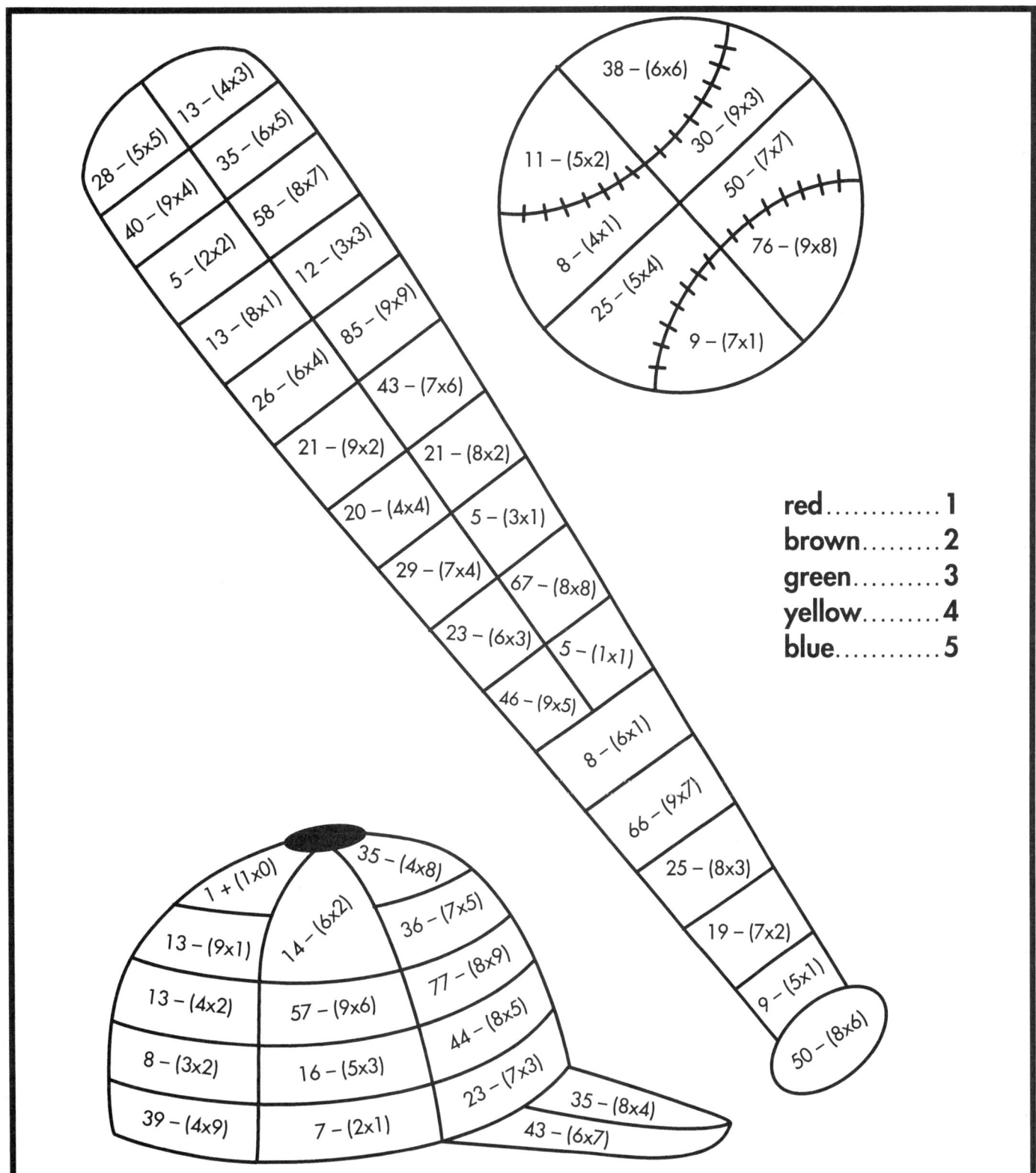

red............1
brown........2
green.........3
yellow........4
blue..........5

Multiplication Puzzles ♦ Activity 3

Correction Key

1. Allow your students to correct their own work.
2. Make a transparency of this puzzle and instruct your students to place the transparency over their completed puzzle for a quick and easy check.

Bear

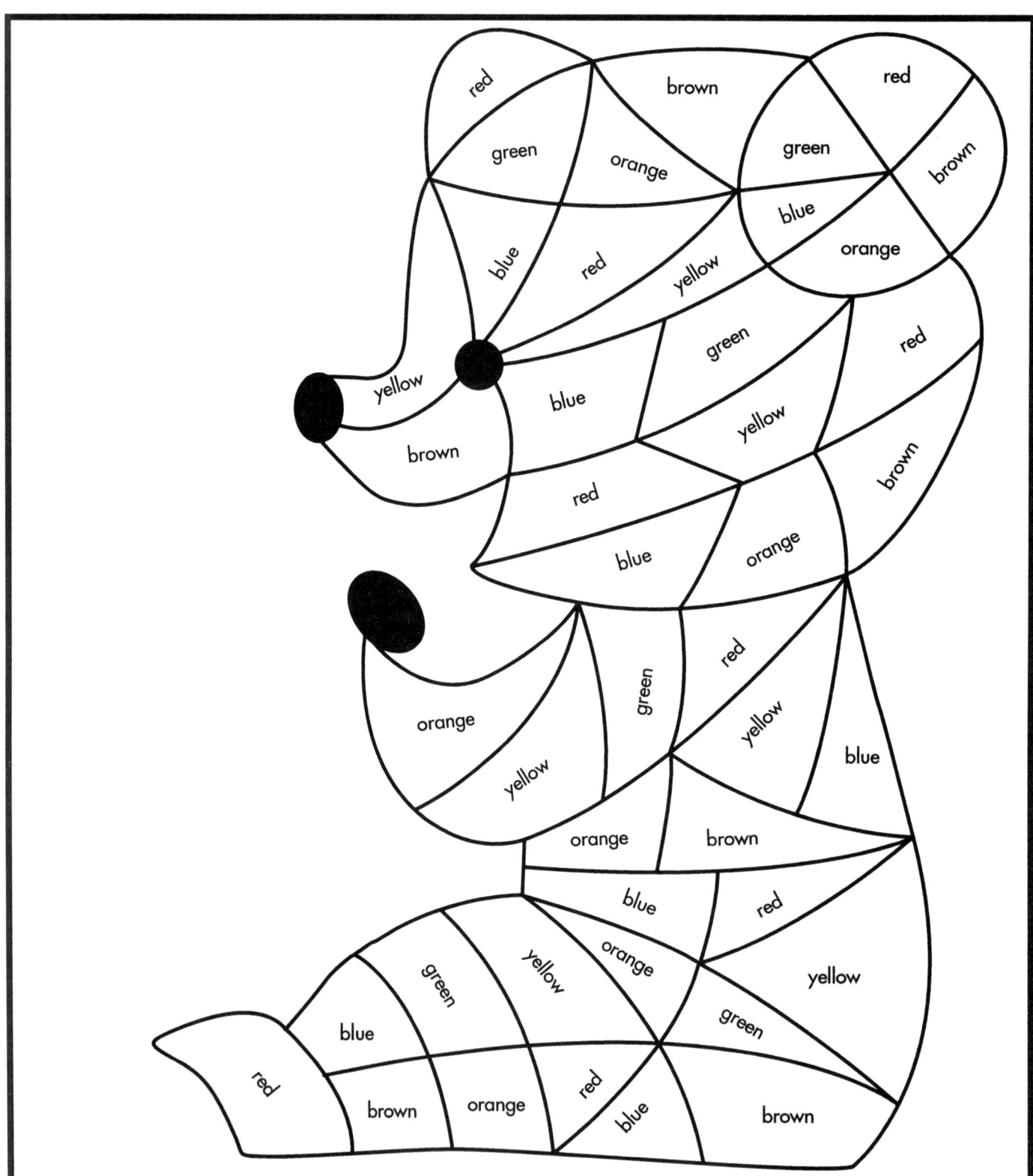

Multiplication Puzzles ✦ **Activity 4**

© Golden Educational Center

Bear

Name _____

Date _____

1. Complete the problems within the parentheses first. Then complete the addition or subtraction problem using your answer.
2. Using your final answer and the color key, color your puzzle correctly.

44 Total Problems

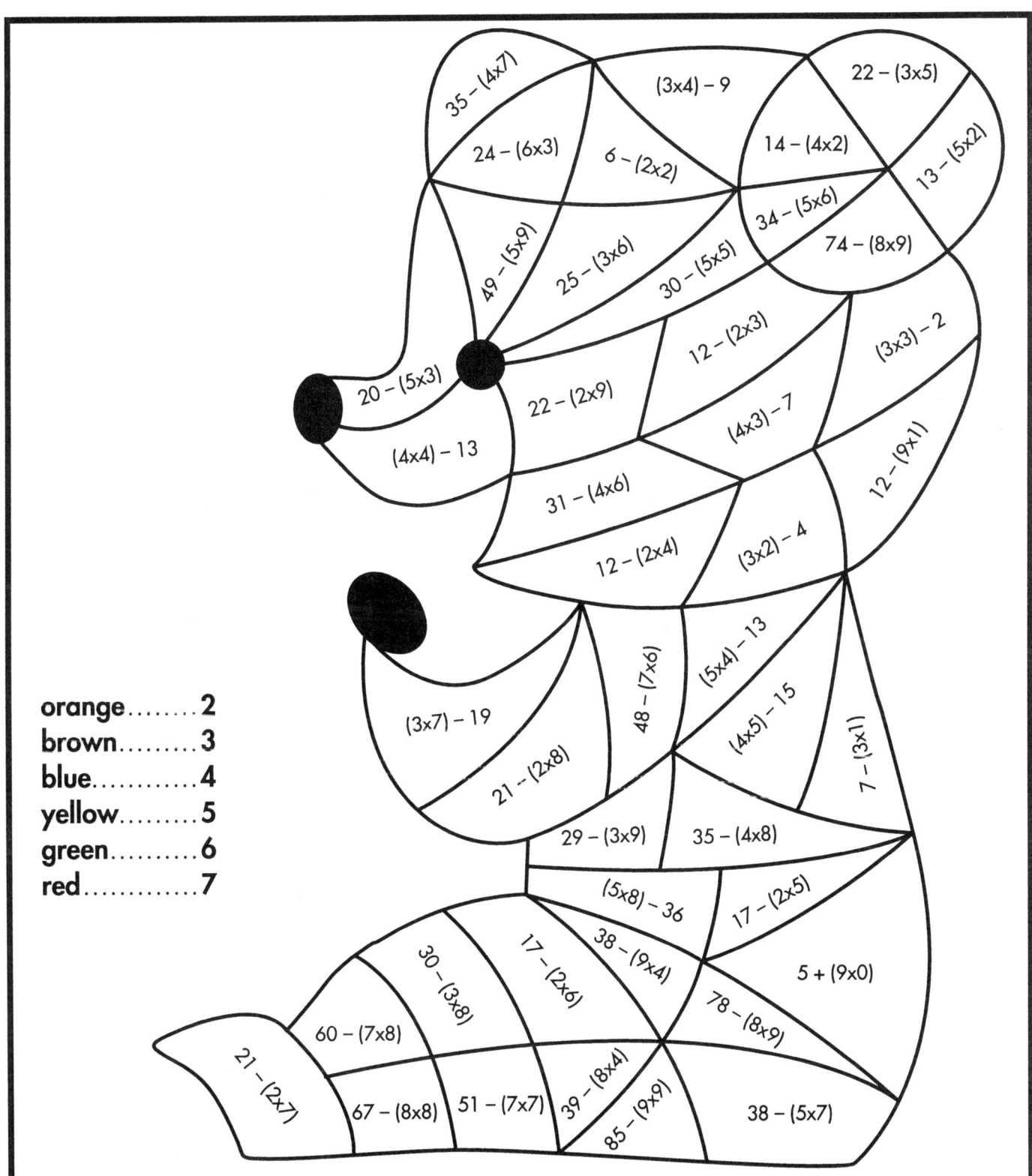

orange 2
brown 3
blue 4
yellow 5
green 6
red 7

Multiplication Puzzles ♦ Activity 4

Correction Key

Bee

1. Allow your students to correct their own work.
2. Make a transparency of this puzzle and instruct your students to place the transparency over their completed puzzle for a quick and easy check.

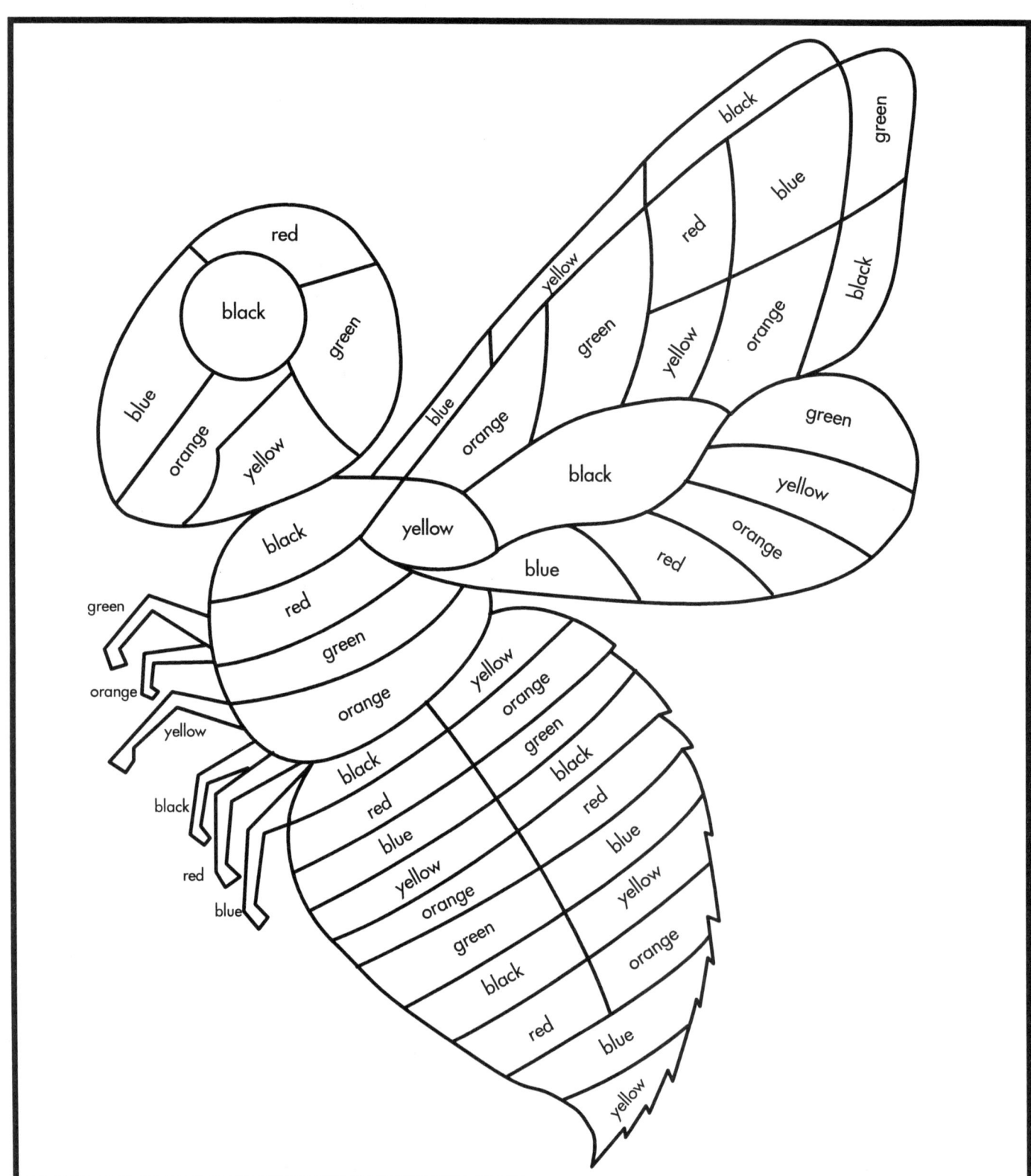

Multiplication Puzzles ✦ Activity 5

Bee

Name _____

Date _____

1. Complete the problems within the parentheses first. Then complete the addition or subtraction problem using your answer.

2. Using your final answer and the color key, color your puzzle correctly.

52 Total Problems

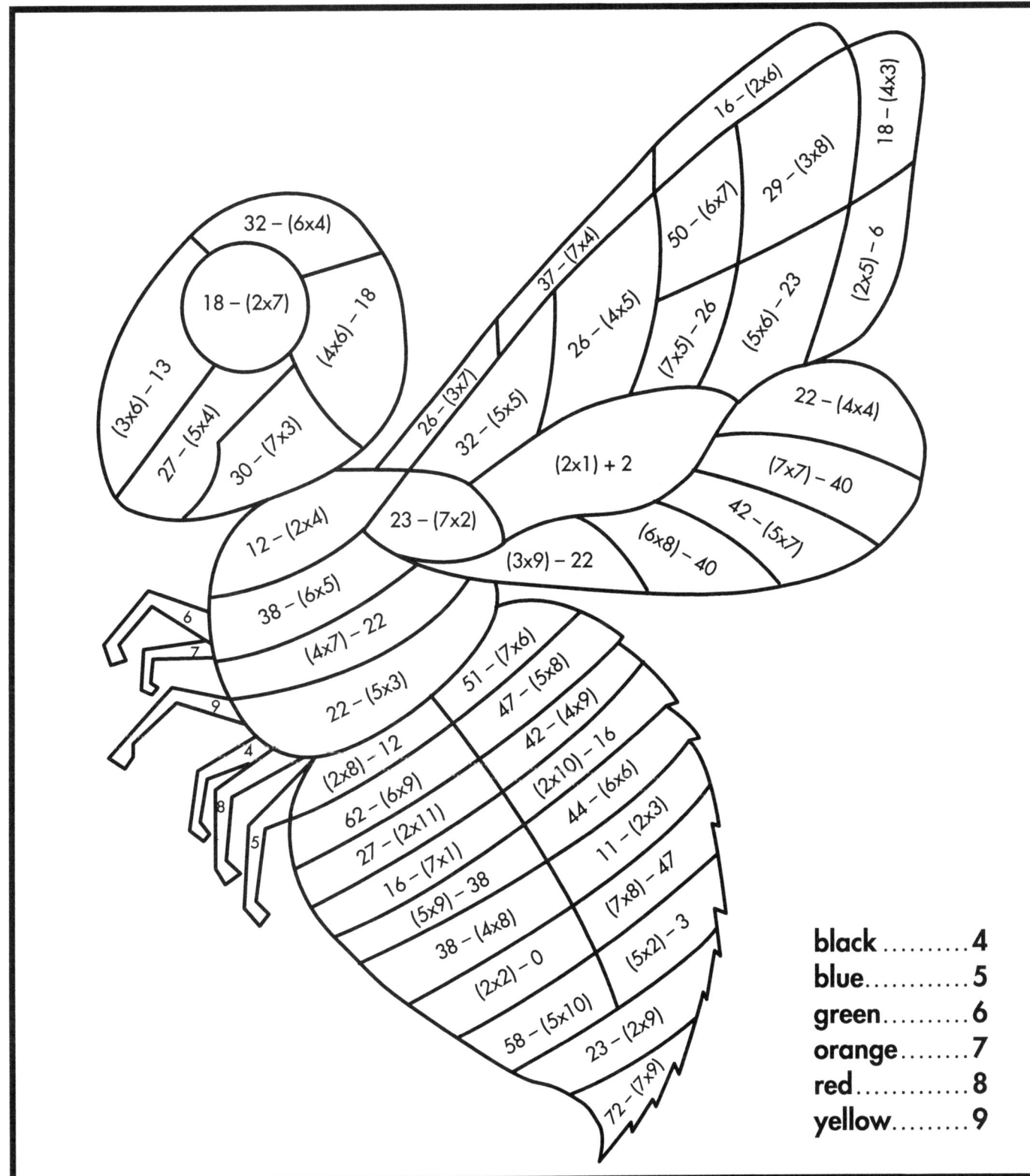

black 4
blue 5
green 6
orange 7
red 8
yellow 9

© Golden Educational Center

Multiplication Puzzles ✦ Activity 5

Correction Key

Bell

1. Allow your students to correct their own work.
2. Make a transparency of this puzzle and instruct your students to place the transparency over their completed puzzle for a quick and easy check.

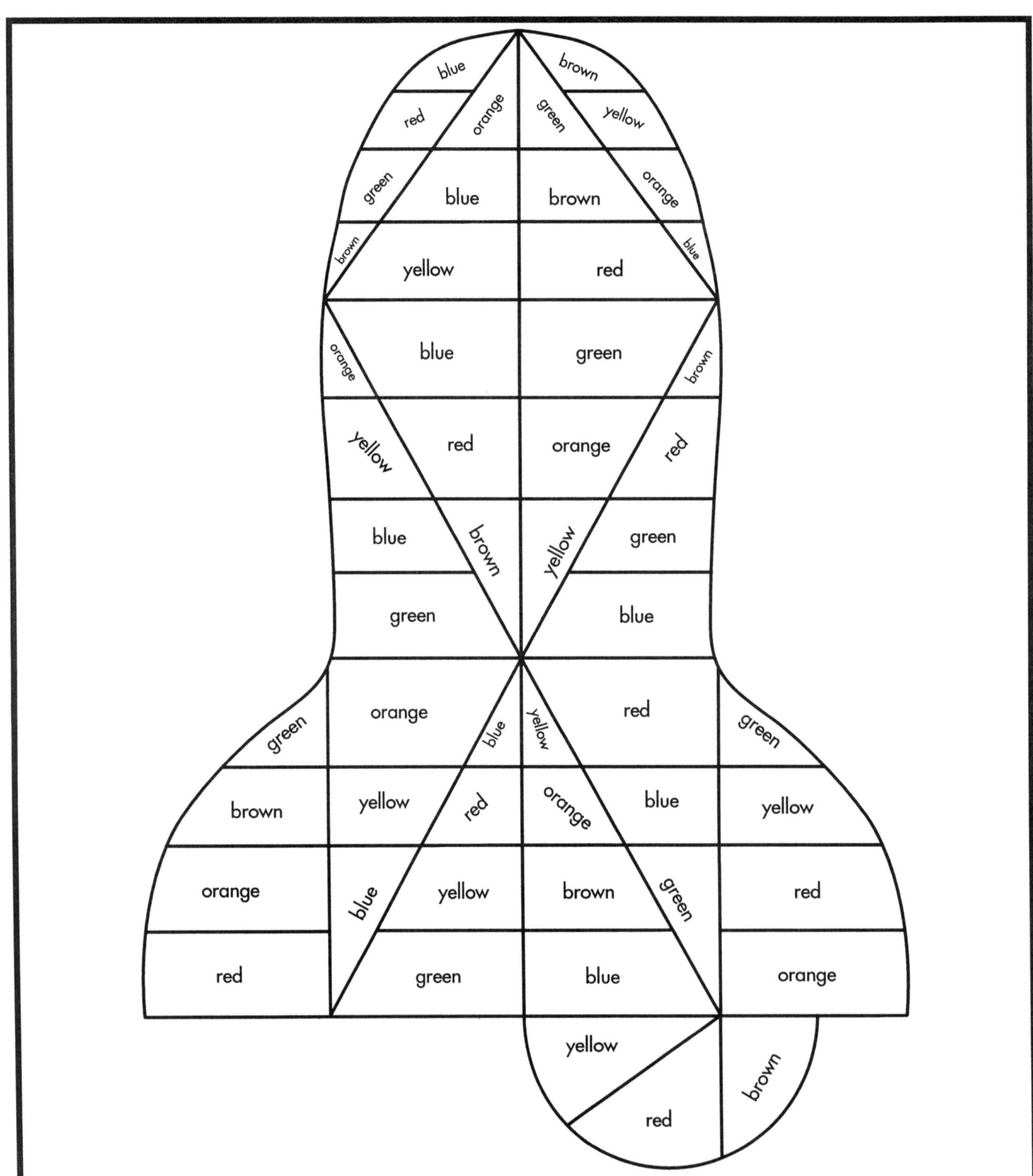

Multiplication Puzzles ✦ Activity 6

Bell

Name _____

Date _____

1. Complete the problems within the parentheses first. Then complete the addition or subtraction problem using your answer.
2. Using your final answer and the color key, color your puzzle correctly.

53 Total Problems

blue............1
brown........2
green..........3
orange........4
red..............5
yellow........6

Puzzle problems:
- 8 − (7x1)
- 66 − (8x8)
- 47 − (7x6)
- 40 − (9x4)
- 30 − (9x3)
- 18 − (4x3)
- 18 − (5x3)
- 7 − (3x2)
- 51 − (7x7)
- 13 − (9x1)
- 65 − (9x7)
- 6 − (6x0)
- 25 − (5x4)
- 10 − (3x3)
- 4 − (7x0)
- 11 − (5x2)
- 39 − (6x6)
- 74 − (9x8)
- 14 − (4x2)
- 61 − (8x7)
- 67 − (7x9)
- 13 − (8x1)
- 36 − (7x5)
- 4 − (2x1)
- 20 − (7x2)
- 28 − (5x5)
- 21 − (9x2)
- 82 − (9x9)
- 39 − (4x9)
- 34 − (6x5)
- 13 − (6x2)
- 6 + (8x0)
- 5 − (5x0)
- 19 − (8x2)
- 30 − (7x4)
- 30 − (8x3)
- 6 − (1x1)
- 49 − (9x5)
- 19 − (6x3)
- 46 − (8x5)
- 4 − (0x3)
- 5 − (4x1)
- 54 − (8x6)
- 5 − (3x1)
- 7 − (2x2)
- 59 − (9x6)
- 11 − (6x1)
- 8 − (5x1)
- 17 − (4x4)
- 28 − (6x4)
- 6 − (9x0)
- 41 − (6x6)
- 34 − (8x4)

© Golden Educational Center

13

Multiplication Puzzles ✦ Activity 6

Correction Key

Bird

1. Allow your students to correct their own work.
2. Make a transparency of this puzzle and instruct your students to place the transparency over their completed puzzle for a quick and easy check.

Multiplication Puzzles ✦ Activity 7

Bird

Name _____

Date _____

1. Complete the problems within the parentheses first. Then complete the addition or subtraction problem using your answer. (You can do the problems on another piece of paper.)
2. Using your final answer and the color key, color your puzzle correctly.

49 Total Problems

orange 2
green 3
yellow 4
red 5
blue 6
brown 7

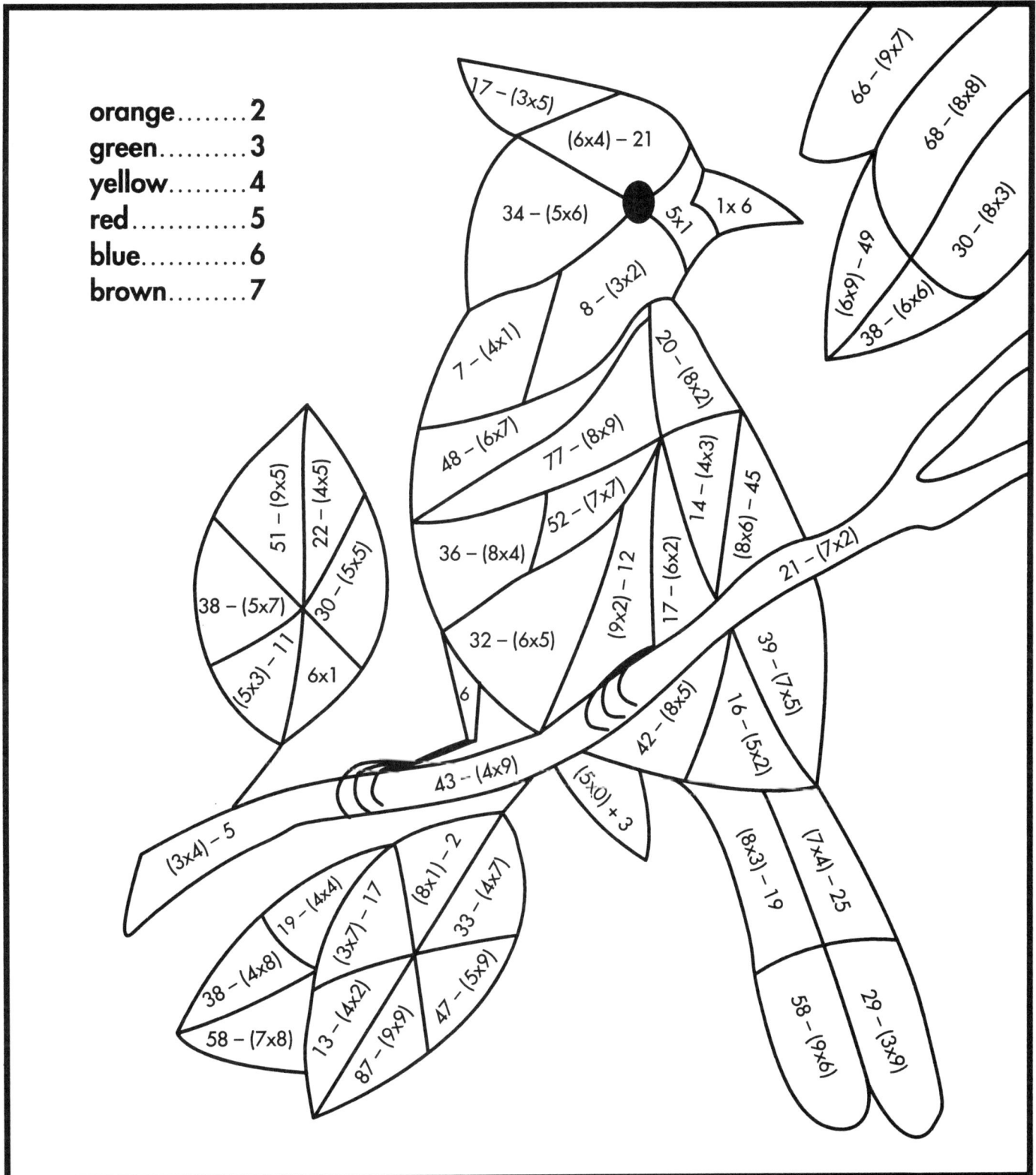

Multiplication Puzzles ✦ Activity 7

Correction Key

Bunny

1. Allow your students to correct their own work.
2. Make a transparency of this puzzle and instruct your students to place the transparency over their completed puzzle for a quick and easy check.

Multiplication Puzzles ✦ Activity 8

© Golden Educational Center

Bunny

Name _____

Date _____

1. Complete the problems within the parentheses first. Then complete the subtraction problem using your answer.

2. Using your final answer and the color key, color your puzzle correctly.

54 Total Problems

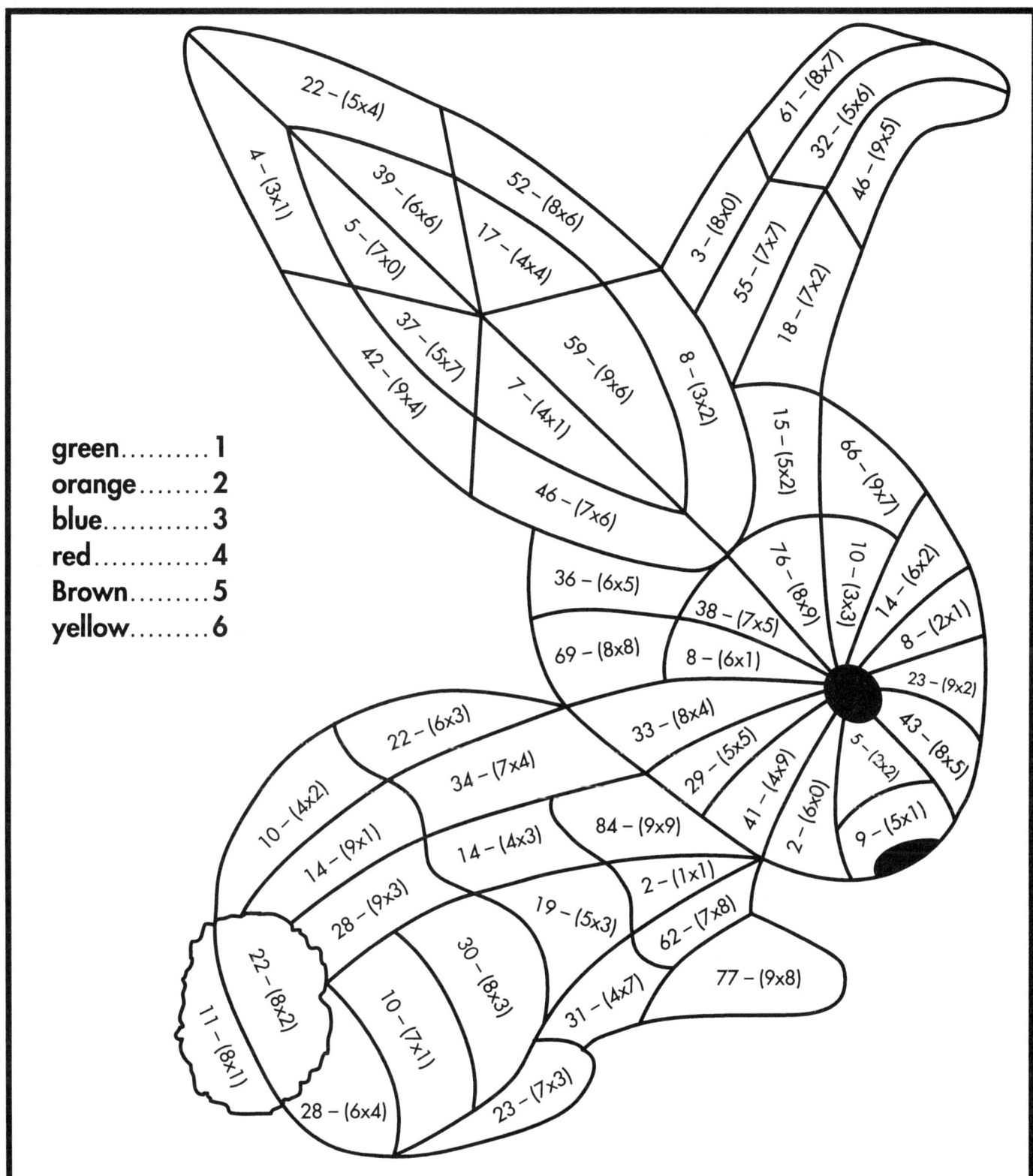

green..........1
orange........2
blue............3
red.............4
Brown.........5
yellow.........6

Multiplication Puzzles ✦ **Activity 8**

Correction Key

1. Allow your students to correct their own work.
2. Make a transparency of this puzzle and instruct your students to place the transparency over their completed puzzle for a quick and easy check.

Butterfly

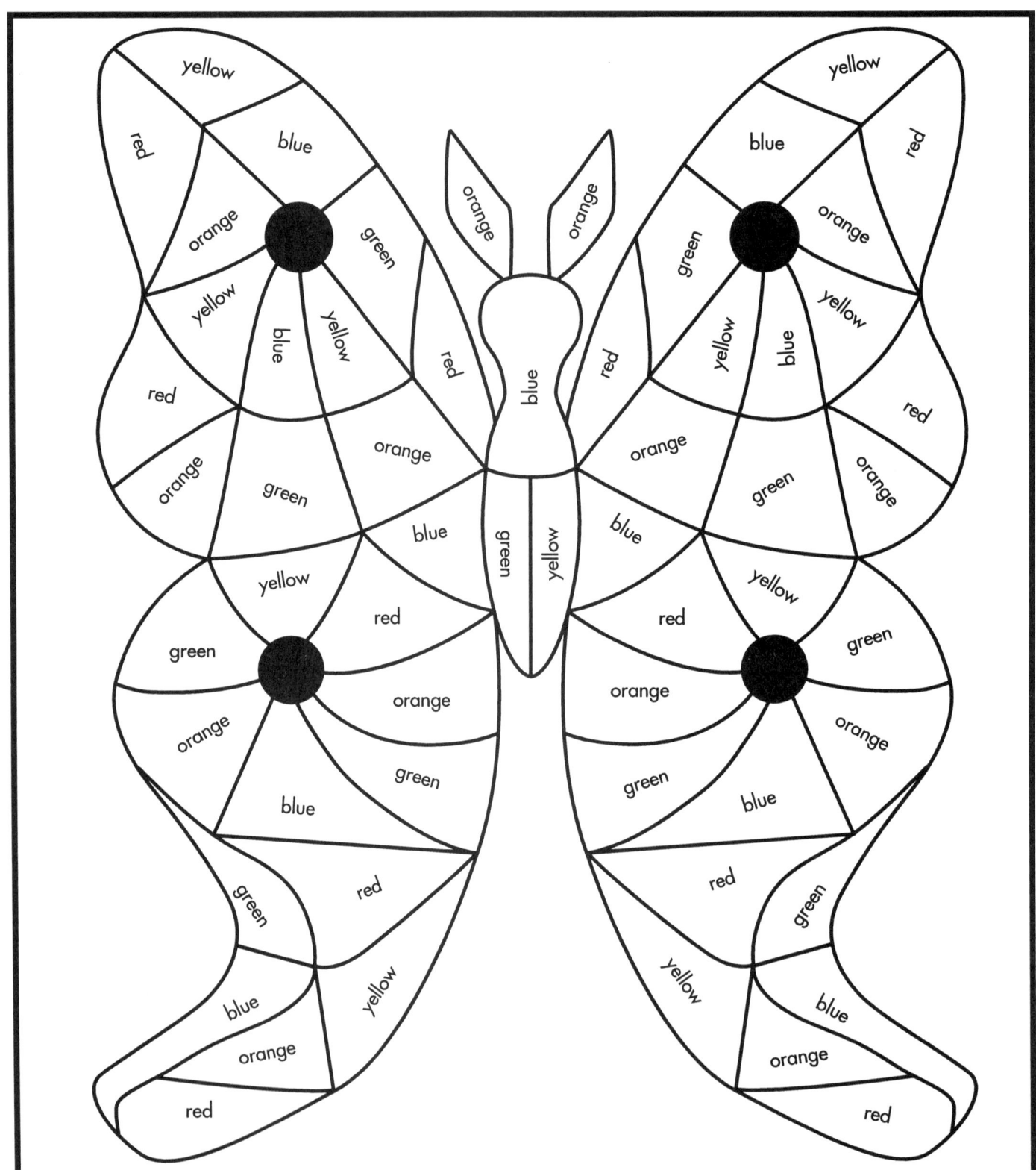

Multiplication Puzzles ✦ Activity 9 © Golden Educational Center

Butterfly

Name _____

Date _____

1. Complete the problems within the parentheses first. Then complete the addition or subtraction problem using your answer.
2. Using your final answer and the color key, color your puzzle correctly.

59 Total Problems

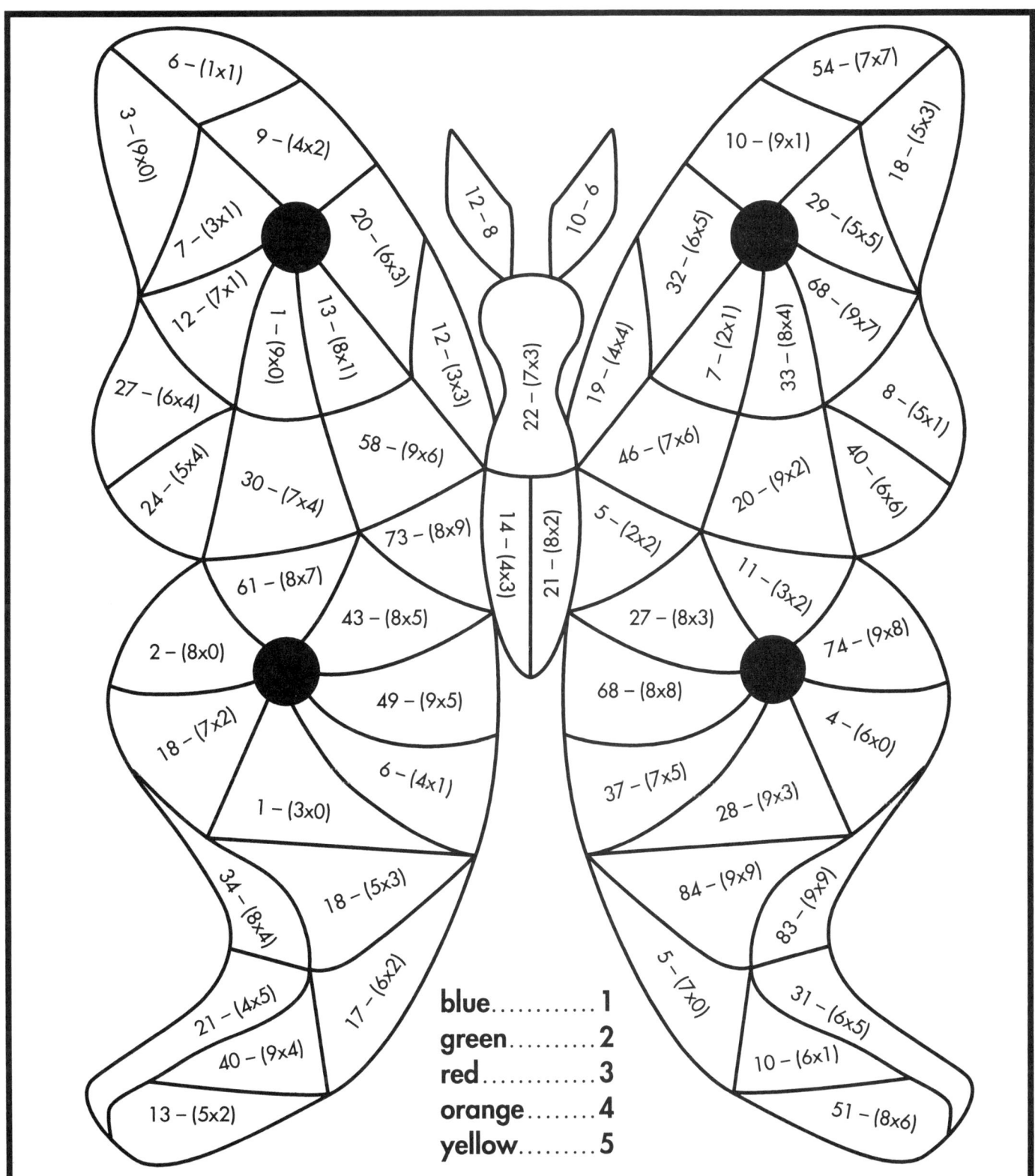

blue.............1
green..........2
red.............3
orange........4
yellow.........5

Multiplication Puzzles ♦ Activity 9

Correction Key

Candles

1. Allow your students to correct their own work.
2. Make a transparency of this puzzle and instruct your students to place the transparency over their completed puzzle for a quick and easy check.

Multiplication Puzzles ✦ Activity 10

© Golden Educational Center

Candles

Name _____

Date _____

1. Complete the problems within the parentheses first. Then complete the subtraction problem using your answer.
2. Using your final answer and the color key, color your puzzle correctly.

64 Total Problems

blue............1
Brown.........2
green..........3
orange........4
red.............5
yellow.........6

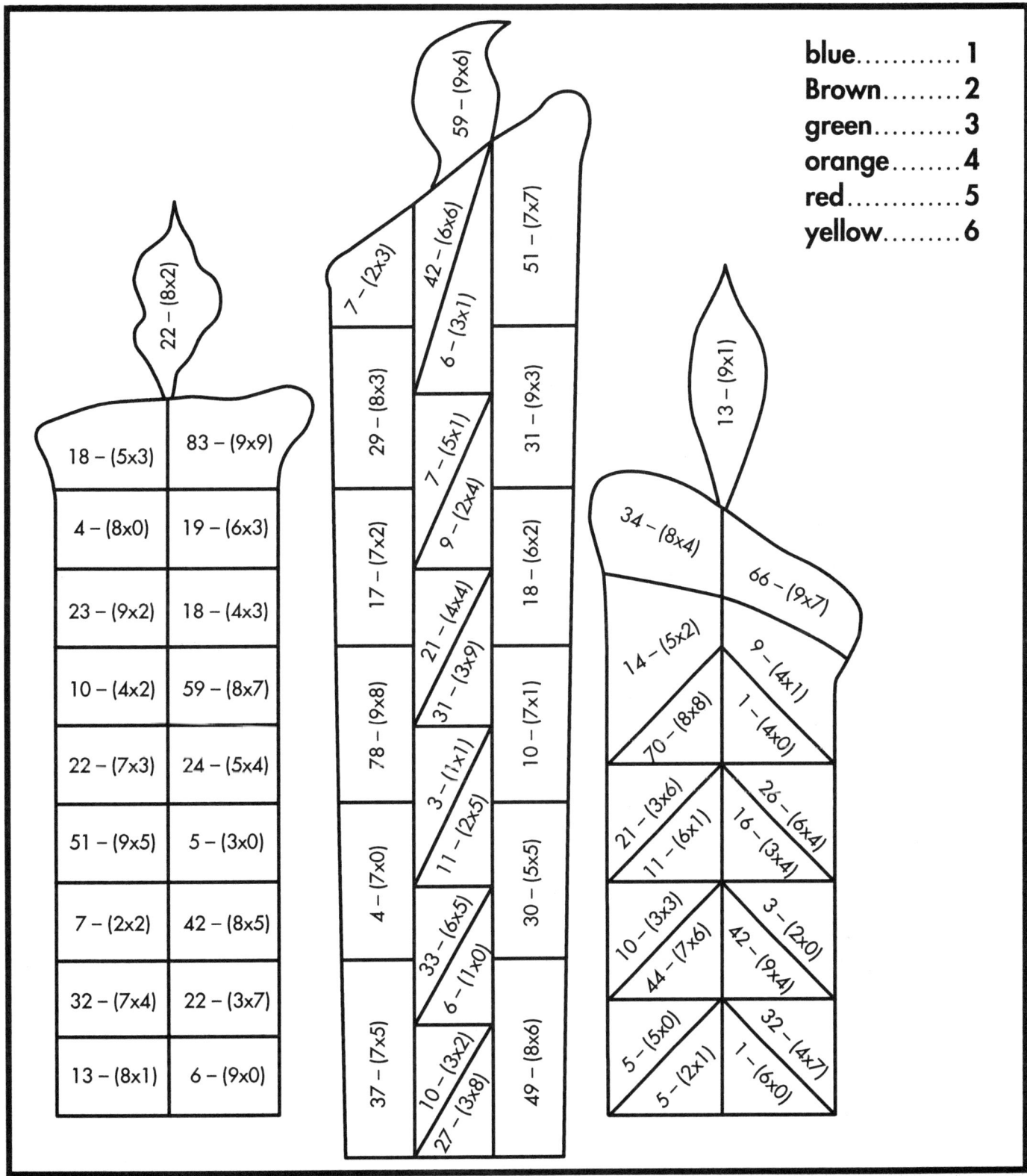

© Golden Educational Center

Multiplication Puzzles ✦ Activity 10

Correction Key **Clipboard**

1. Allow your students to correct their own work.
2. Make a transparency of this puzzle and instruct your students to place the transparency over their completed puzzle for a quick and easy check.

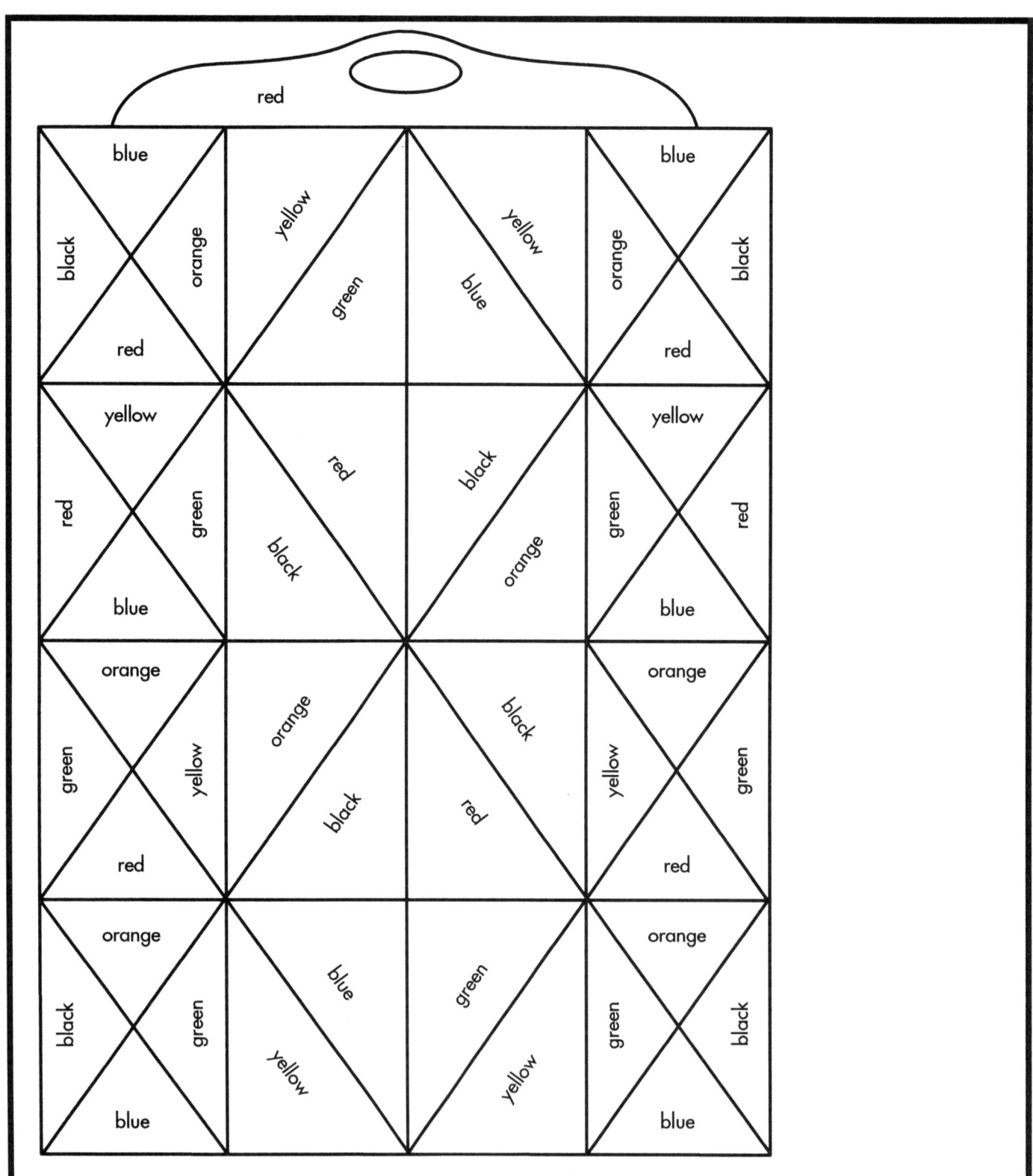

Multiplication Puzzles ✦ Activity 11

Clipboard

Name _____

Date _____

1. Complete the problems within the parentheses first. Then complete the subtraction problem using your answer. (You can do the problems on another piece of paper.)

2. Using your final answer and the color key, color your puzzle correctly.

49 Total Problems

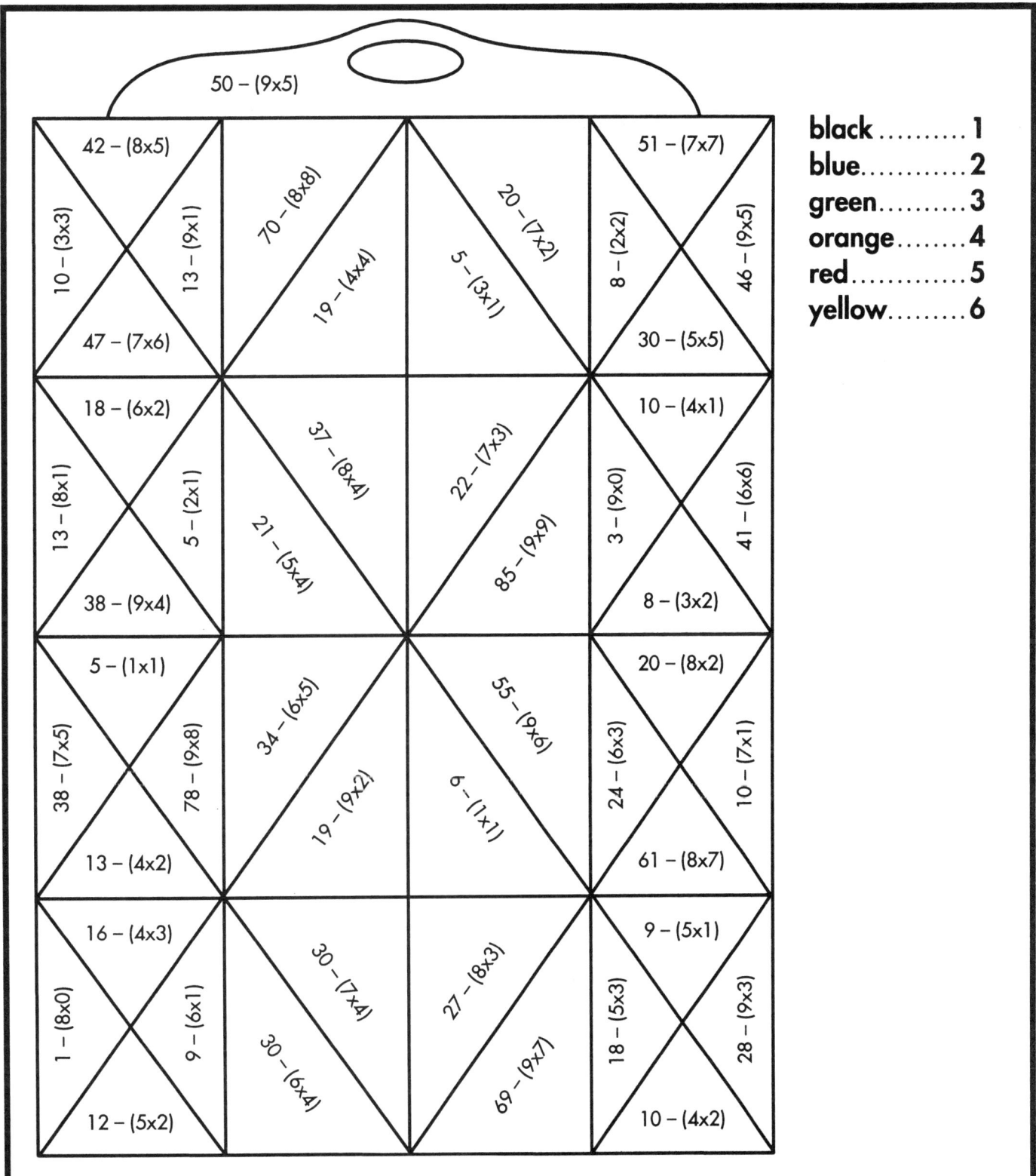

black 1
blue 2
green 3
orange 4
red 5
yellow 6

Multiplication Puzzles ✦ Activity 11

Correction Key

Cosmos Flower

1. Allow your students to correct their own work.
2. Make a transparency of this puzzle and instruct your students to place the transparency over their completed puzzle for a quick and easy check.

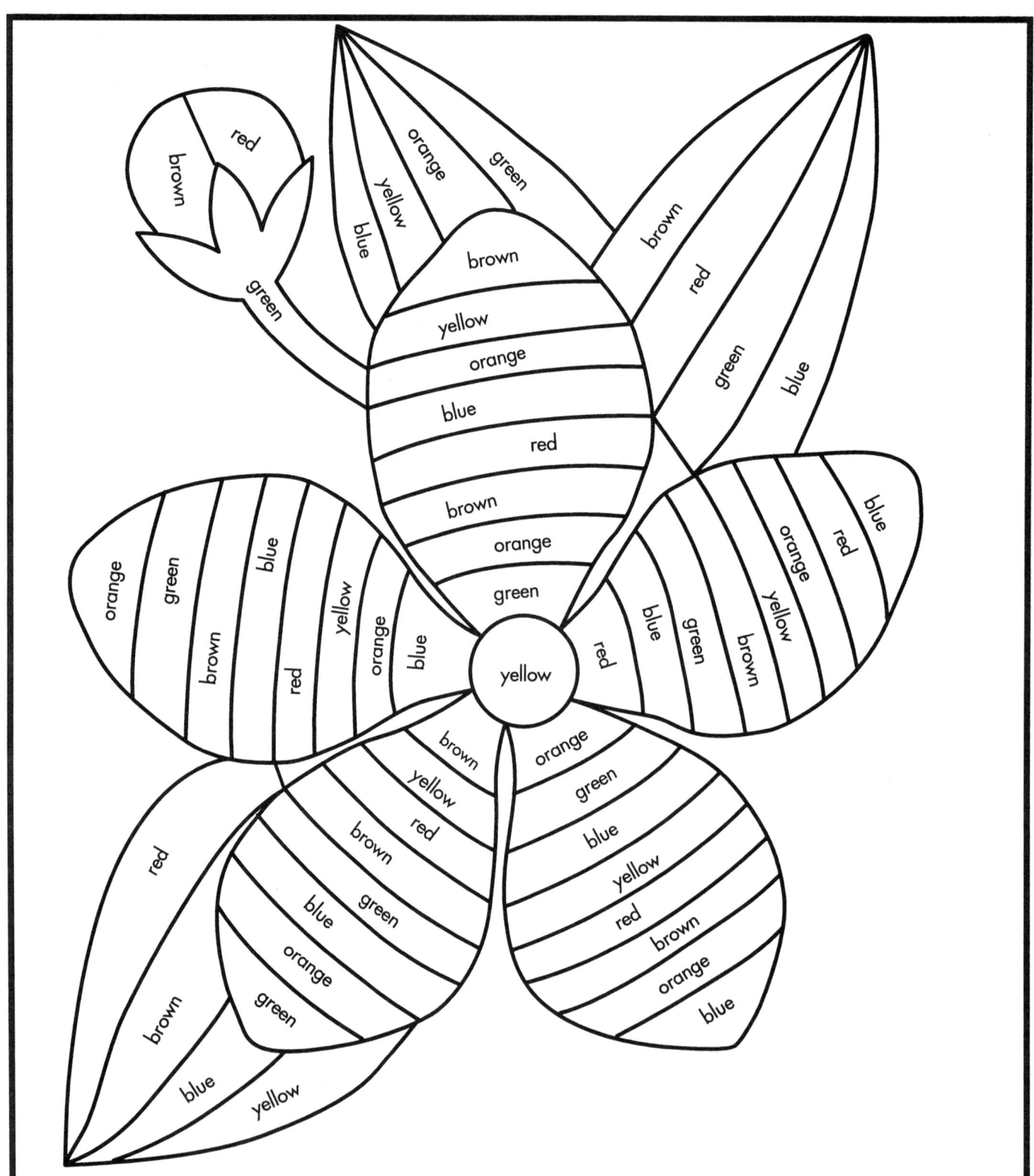

Multiplication Puzzles ✦ Activity 12

Cosmos Flower

Name _____

Date _____

1. Complete the problems within the parentheses first. Then complete the subtraction problem using your answer.
2. Using your final answer and the color key, color your puzzle correctly.

55 Total Problems

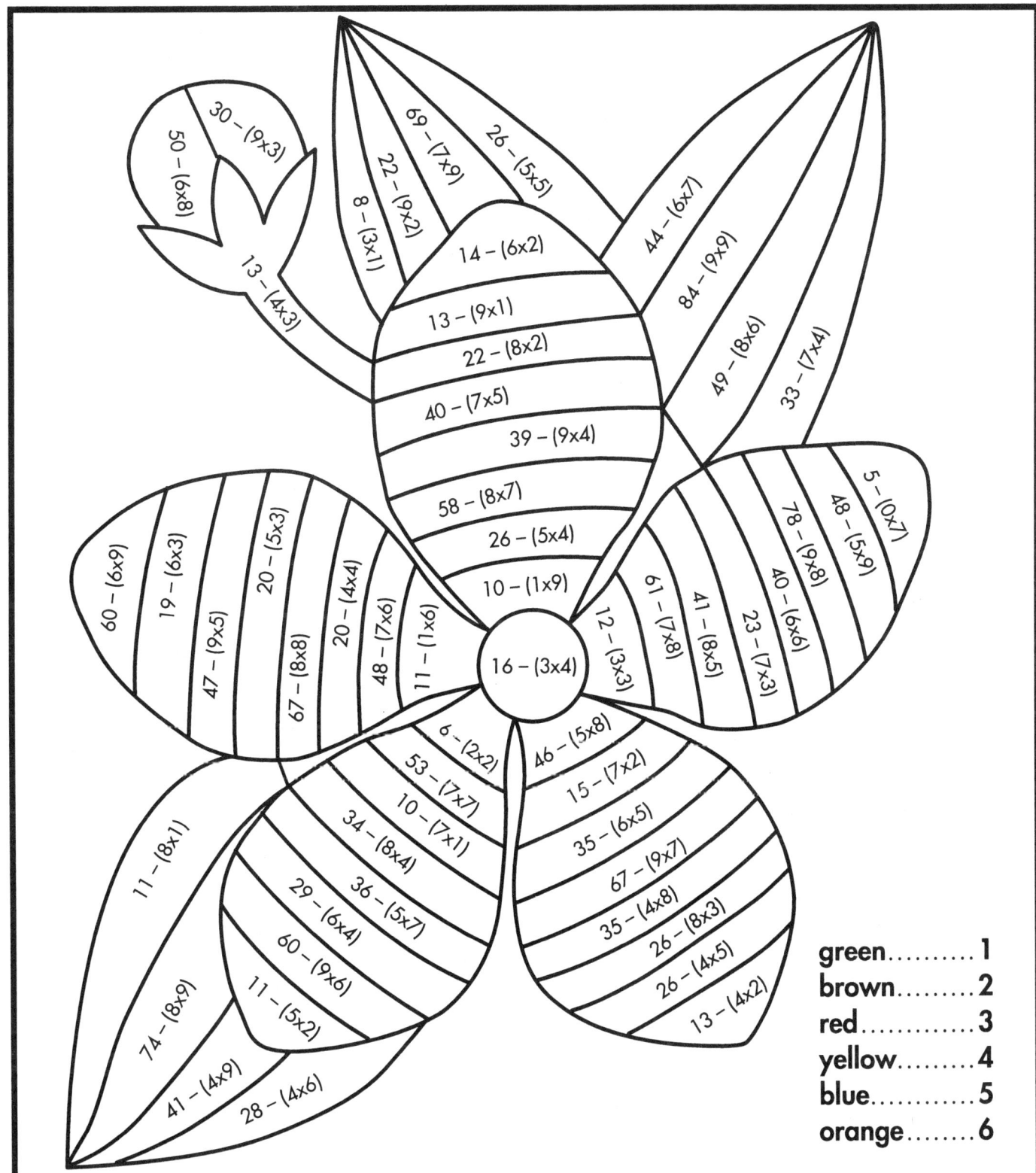

green..........1
brown.........2
red.............3
yellow.........4
blue............5
orange........6

© Golden Educational Center

Multiplication Puzzles ✦ Activity 12

Correction Key

Cupid

1. Allow your students to correct their own work.
2. Make a transparency of this puzzle and instruct your students to place the transparency over their completed puzzle for a quick and easy check.

Multiplication Puzzles ✦ Activity 13

Cupid

1. Complete the problems within the parentheses first. Then complete the subtraction problem using your answer.
2. Using your final answer and the color key, color your puzzle correctly.

58 Total Problems

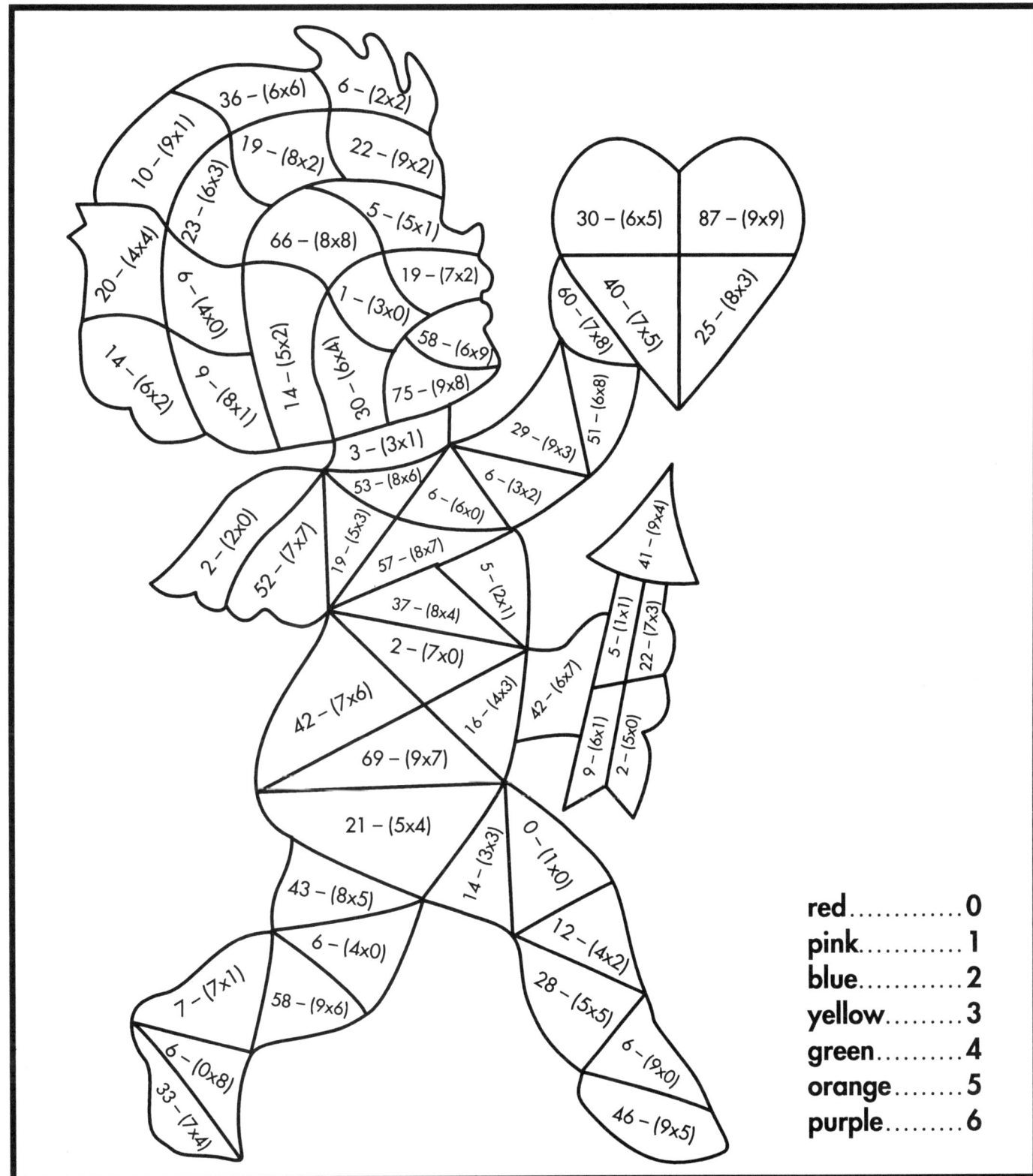

red............0
pink...........1
blue...........2
yellow.........3
green..........4
orange........5
purple.........6

Multiplication Puzzles ✦ Activity 13

Correction Key

Daffodil

1. Allow your students to correct their own work.
2. Make a transparency of this puzzle and instruct your students to place the transparency over their completed puzzle for a quick and easy check.

Multiplication Puzzles ✦ Activity 14

Daffodil

Name _____
Date _____

1. Complete the problems within the parentheses first. Then complete the addition or subtraction problem using your answer.
2. Using your final answer and the color key, color your puzzle correctly.

53 Total Problems

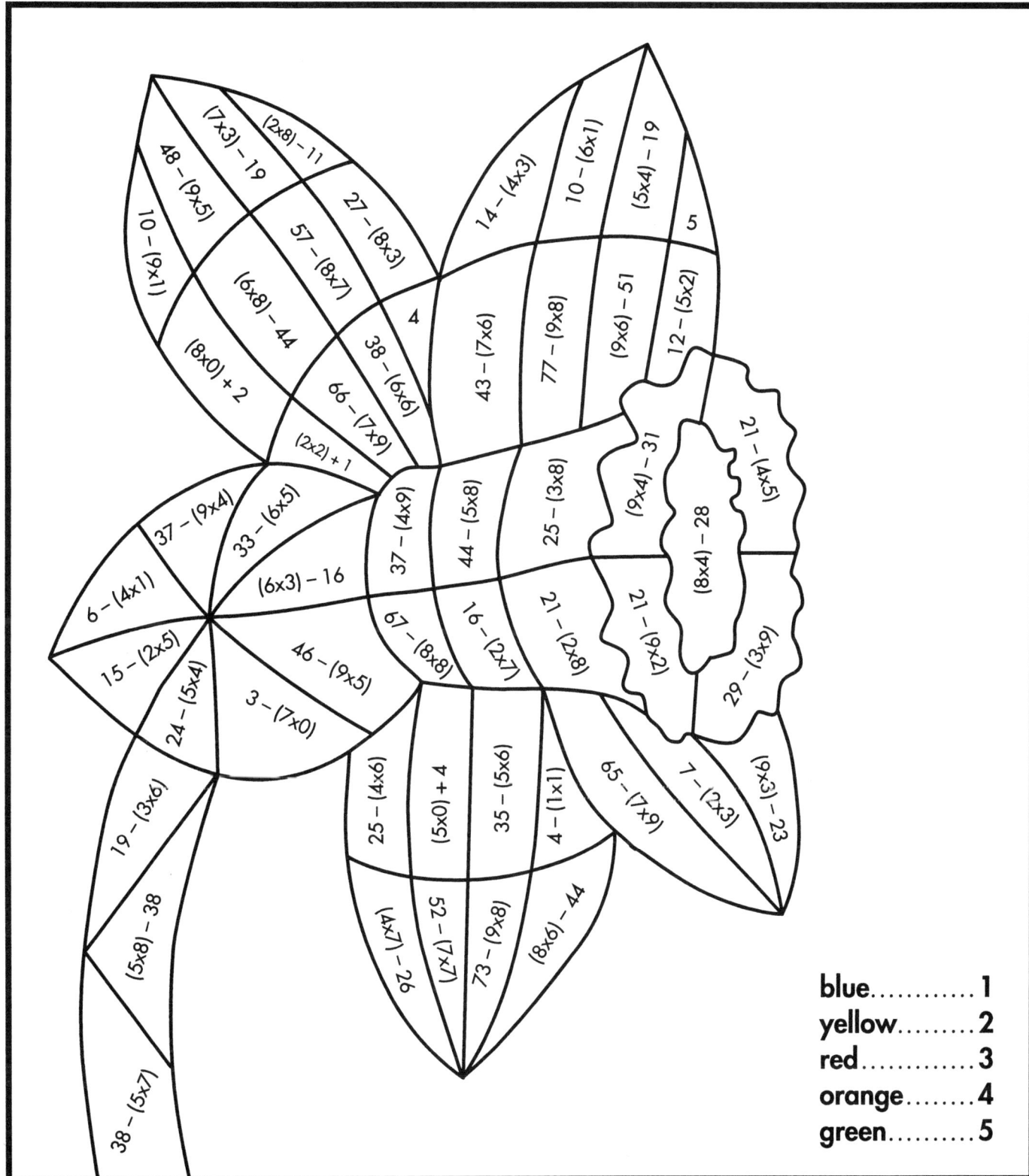

blue............1
yellow.........2
red............3
orange........4
green.........5

Correction Key

Elephant

1. Allow your students to correct their own work.
2. Make a transparency of this puzzle and instruct your students to place the transparency over their completed puzzle for a quick and easy check.

Multiplication Puzzles ✦ Activity 15

Elephant

Name _____

Date _____

1. Complete the problems within the parentheses first. Then complete the subtraction problem using your answer.

2. Using your final answer and the color key, color your puzzle correctly.

38 Total Problems

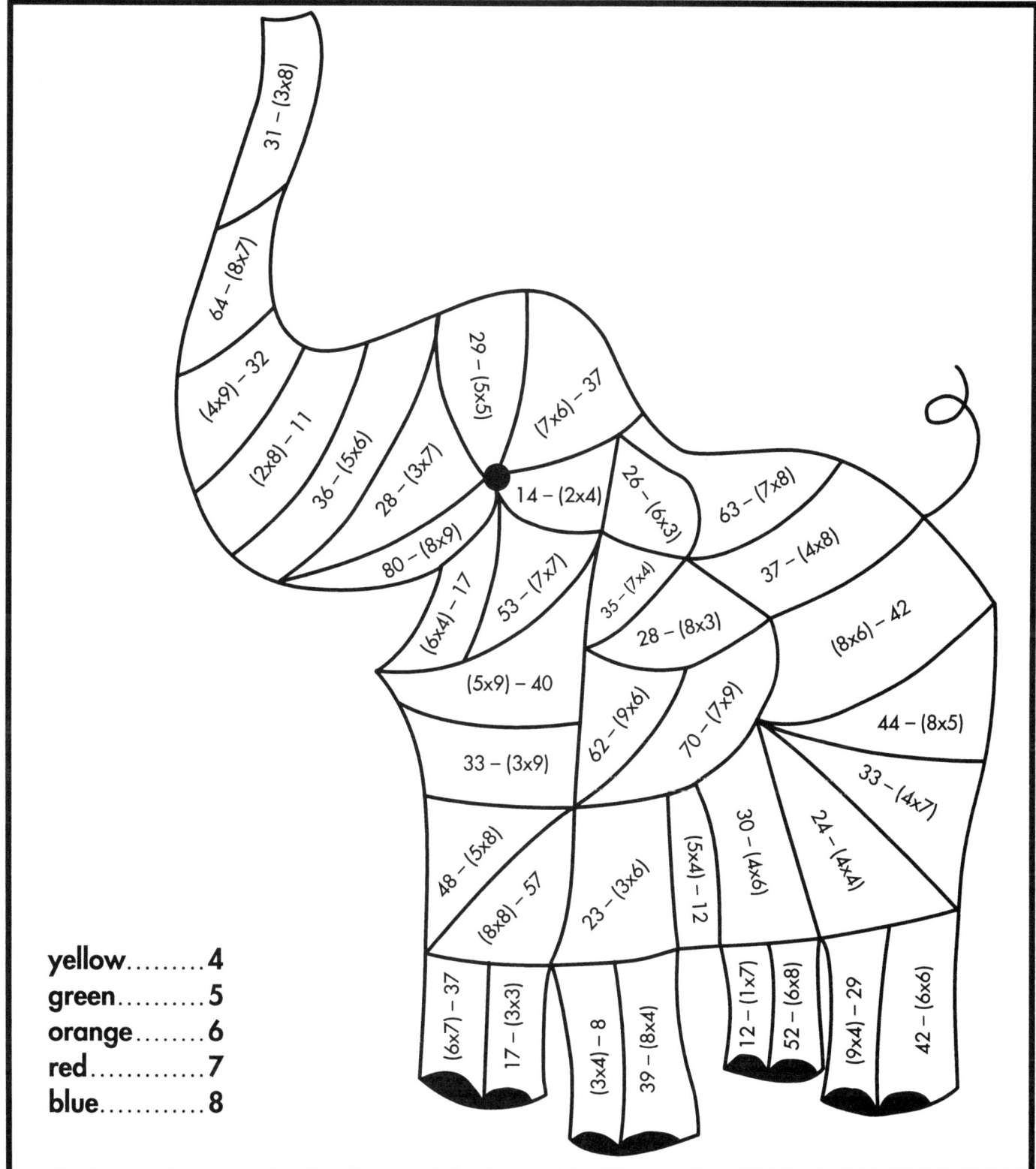

yellow.........4
green..........5
orange........6
red............7
blue............8

Multiplication Puzzles ✦ Activity 15

Correction Key

Fish

1. Allow your students to correct their own work.
2. Make a transparency of this puzzle and instruct your students to place the transparency over their completed puzzle for a quick and easy check.

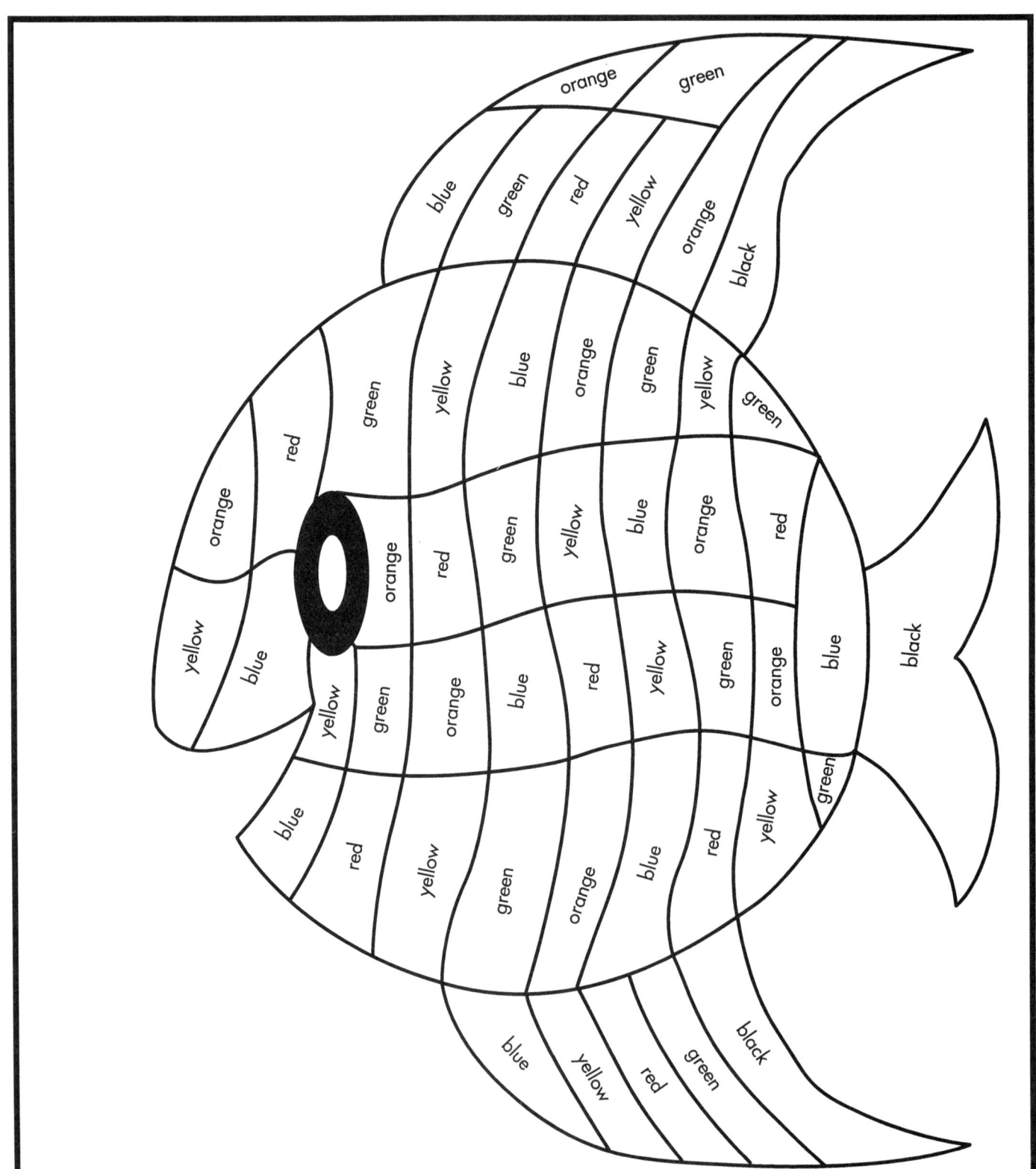

Multiplication Puzzles ✦ Activity 16

Fish

Name _____

Date _____

1. Complete the problems within the parentheses first. Then complete the subtraction problem using your answer.

2. Using your final answer and the color key, color your puzzle correctly.

51 Total Problems

yellow........2
green.........3
blue............4
red.............5
orange.......6
black..........7

Multiplication Puzzles ✦ Activity 16

Correction Key

Flag

1. Allow your students to correct their own work.
2. Make a transparency of this puzzle and instruct your students to place the transparency over their completed puzzle for a quick and easy check.

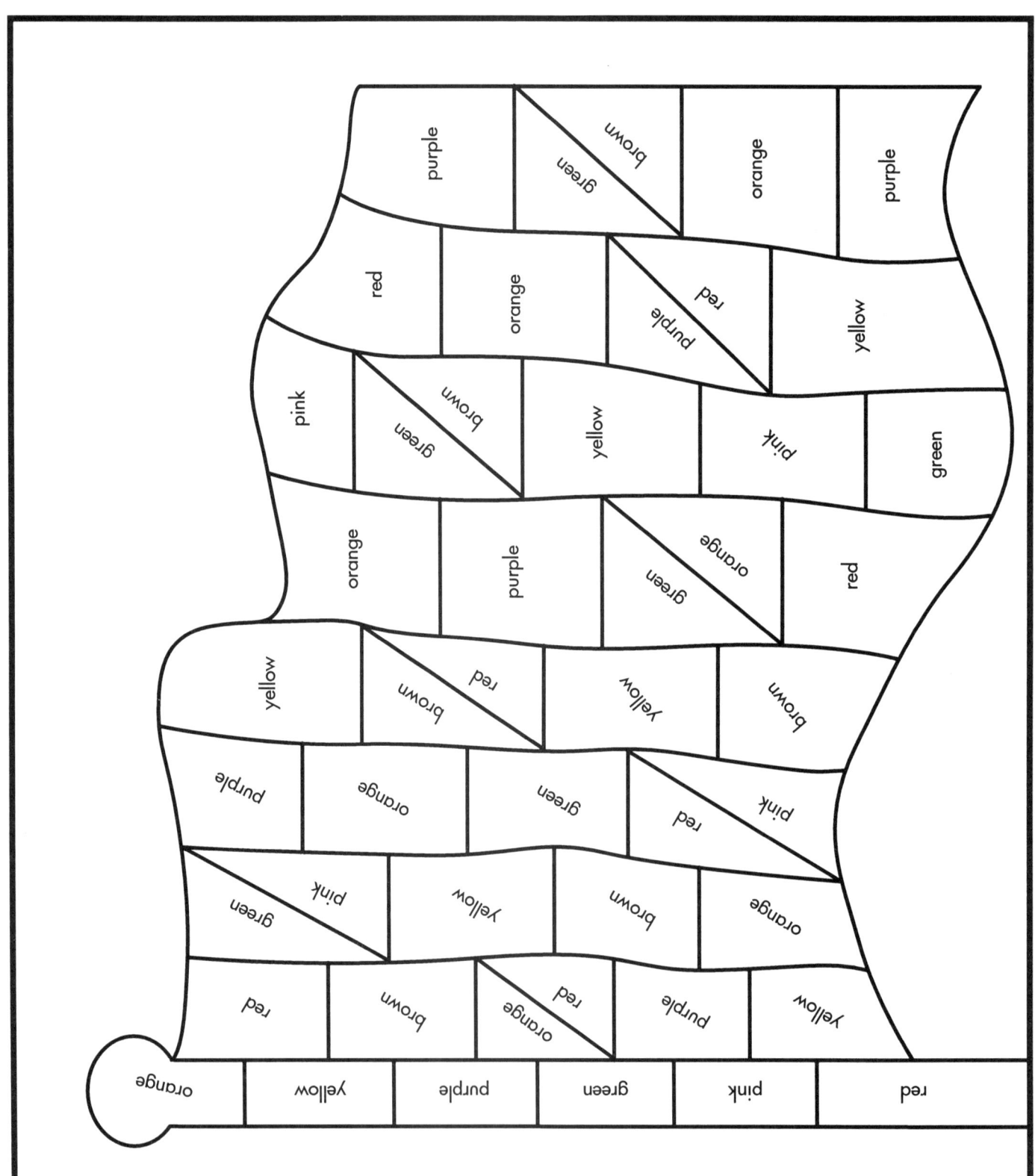

Multiplication Puzzles ♦ Activity 17

Flag

Name _____

Date _____

1. Complete the problems within the parentheses first. Then complete the subtraction problem using your answer. (You can do the problems on another piece of paper.)
2. Using your final answer and the color key, color your puzzle correctly.

48 Total Problems

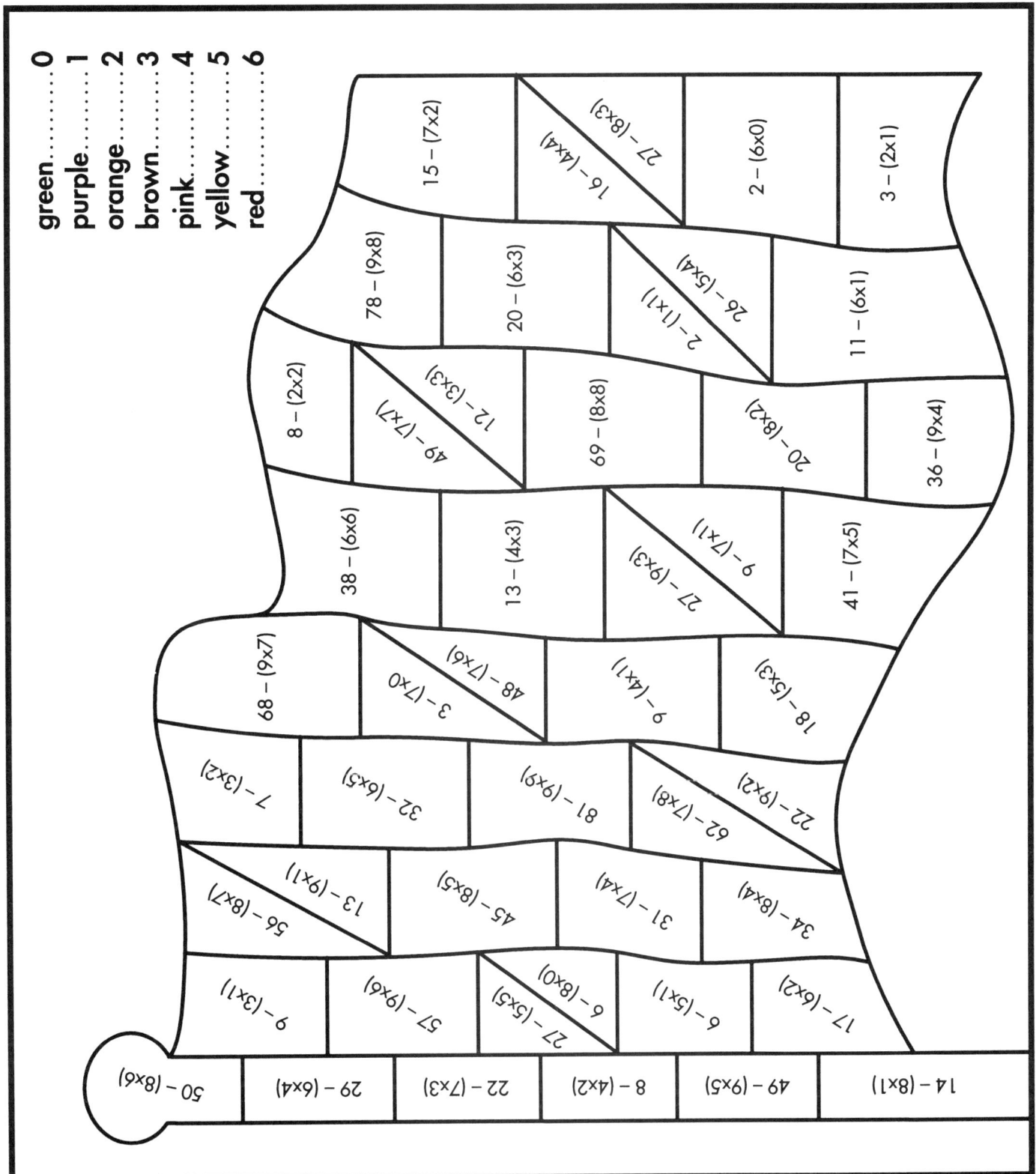

Multiplication Puzzles ✦ Activity 17

Correction Key

Frog

1. Allow your students to correct their own work.
2. Make a transparency of this puzzle and instruct your students to place the transparency over their completed puzzle for a quick and easy check.

Multiplication Puzzles ✦ Activity 18

Frog

Name _____

Date _____

1. Complete the problems within the parentheses first. Then complete the subtraction problem using your answer. (You can do the problems on another piece of paper.)
2. Using your final answer and the color key, color your puzzle correctly.

51 Total Problems

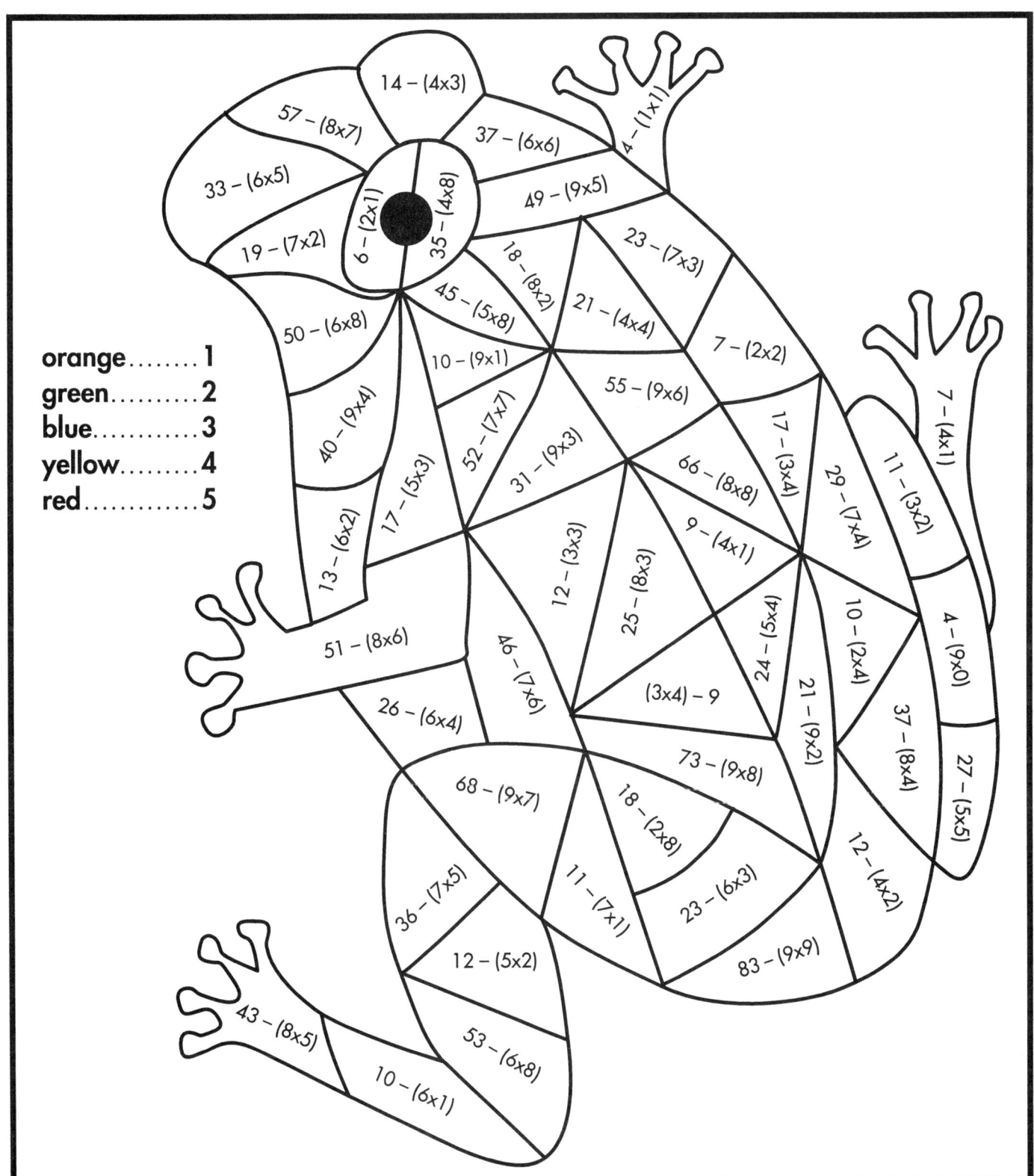

orange........1
green.........2
blue...........3
yellow........4
red............5

© Golden Educational Center

37

Multiplication Puzzles ✦ Activity 18

Correction Key

George Washington

1. Allow your students to correct their own work.
2. Make a transparency of this puzzle and instruct your students to place the transparency over their completed puzzle for a quick and easy check.

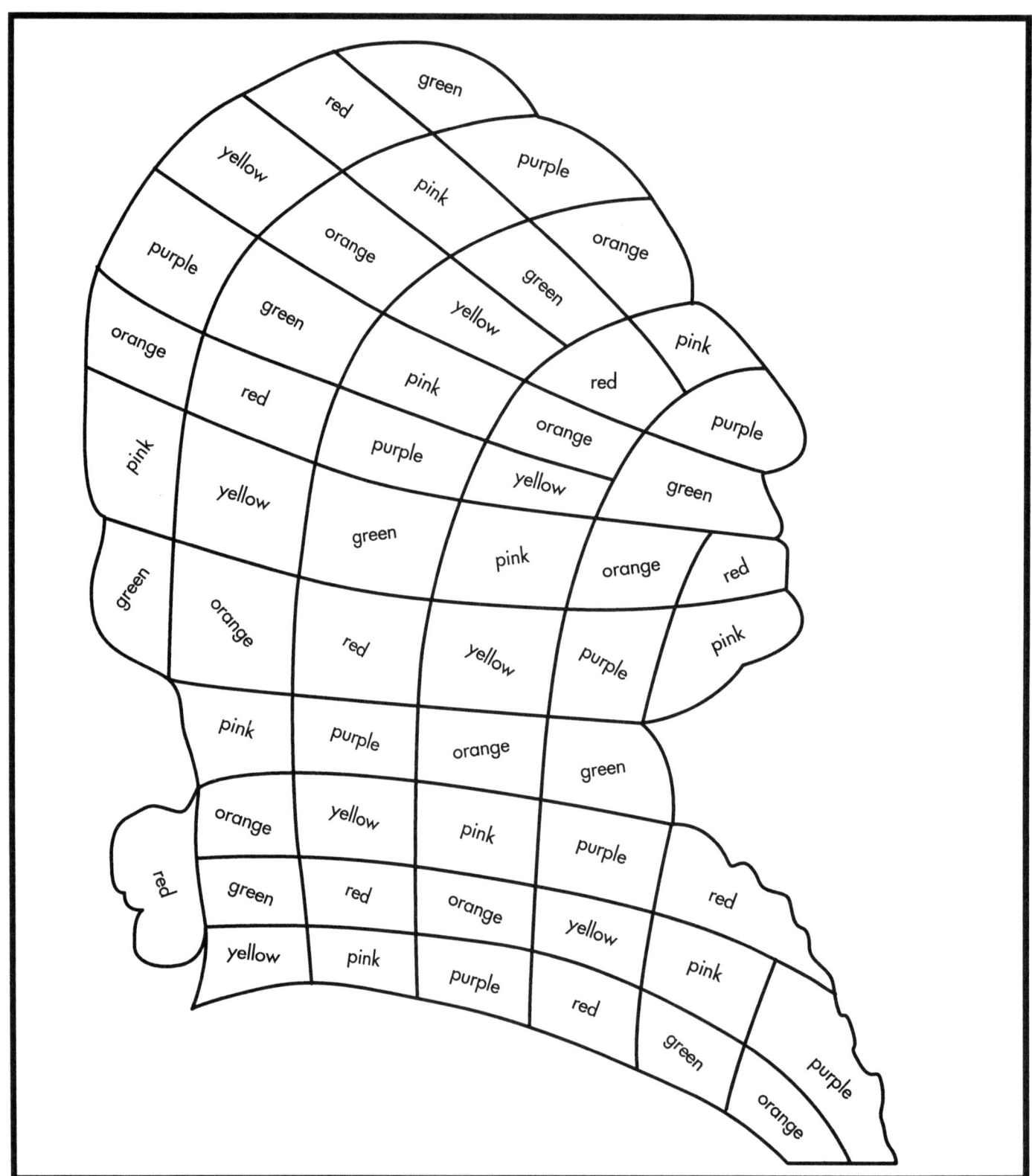

Multiplication Puzzles ✦ Activity 19

George Washington

Name _____

Date _____

1. Complete the problems within the parentheses first. Then complete the subtraction problem using your answer. (You can do the problems on another piece of paper.)
2. Using your final answer and the color key, color your puzzle correctly.

55 Total Problems

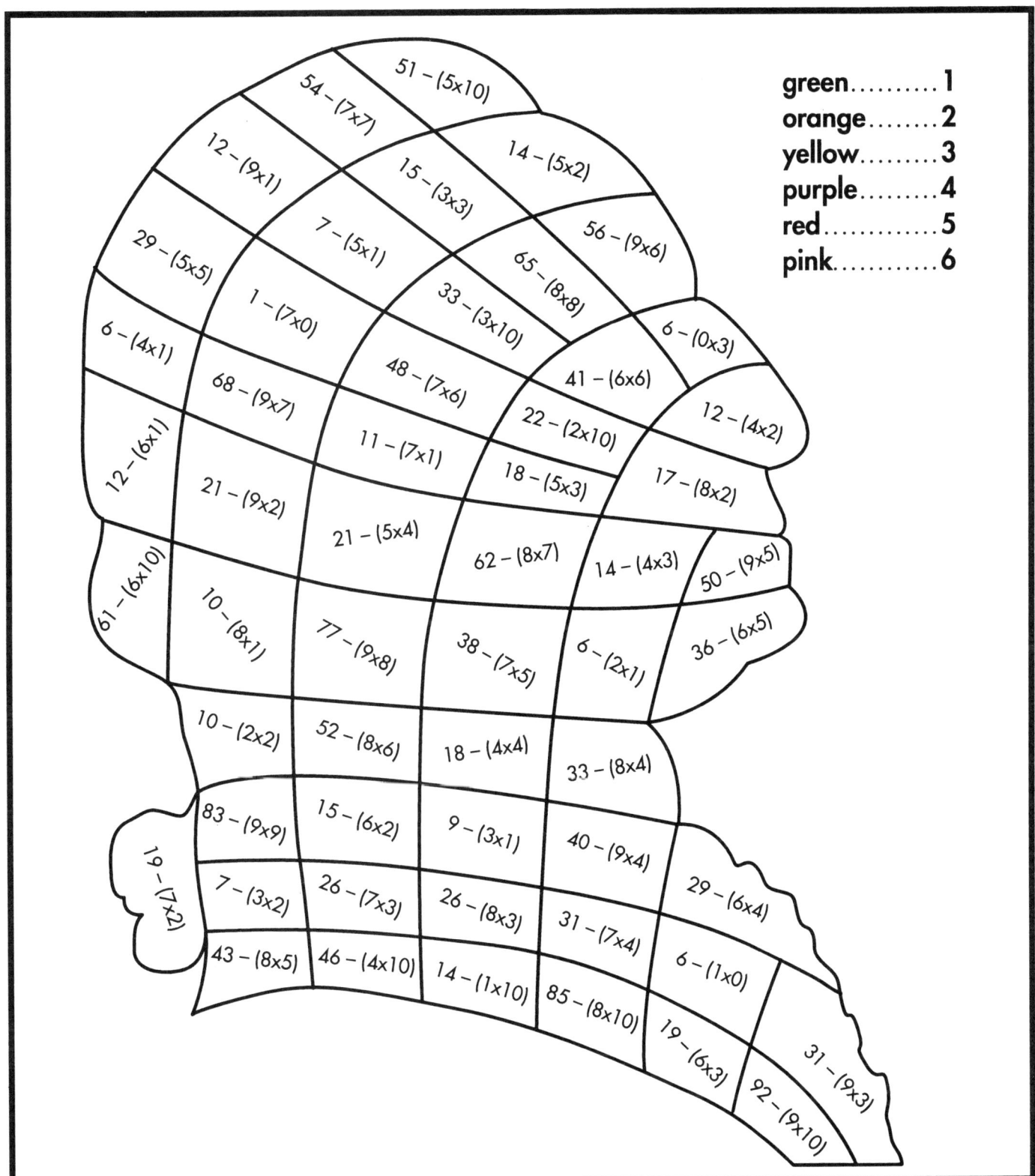

green..........1
orange........2
yellow........3
purple........4
red............5
pink...........6

Multiplication Puzzles ✦ Activity 19

Correction Key

Groundhog

1. Allow your students to correct their own work.
2. Make a transparency of this puzzle and instruct your students to place the transparency over their completed puzzle for a quick and easy check.

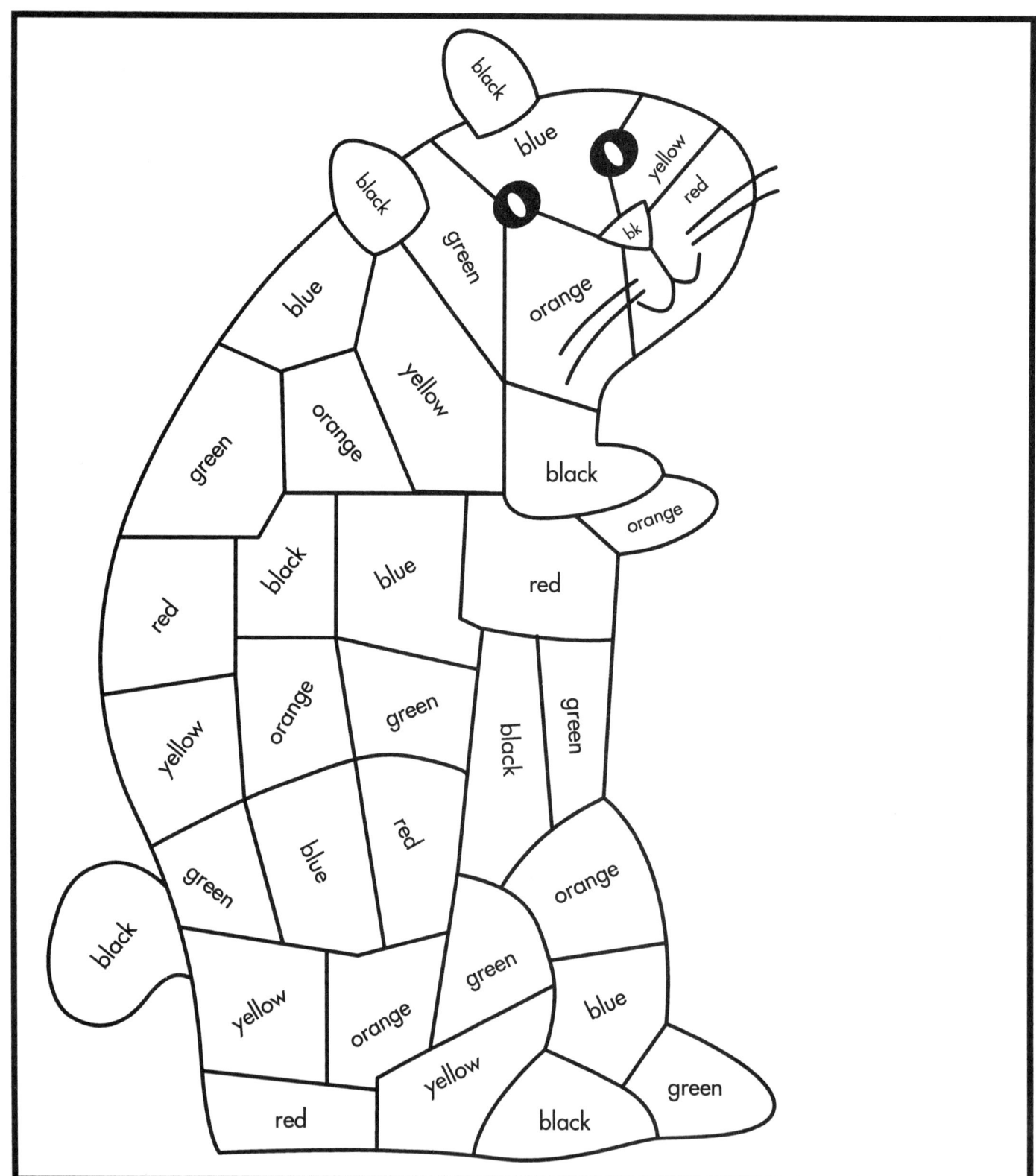

Multiplication Puzzles ✦ Activity 20

Groundhog

Name _____

Date _____

1. Complete the problems within the parentheses first. Then complete the subtraction problem using your answer. (You can do the problems on another piece of paper.)

2. Using your final answer and the color key, color your puzzle correctly.

36 Total Problems

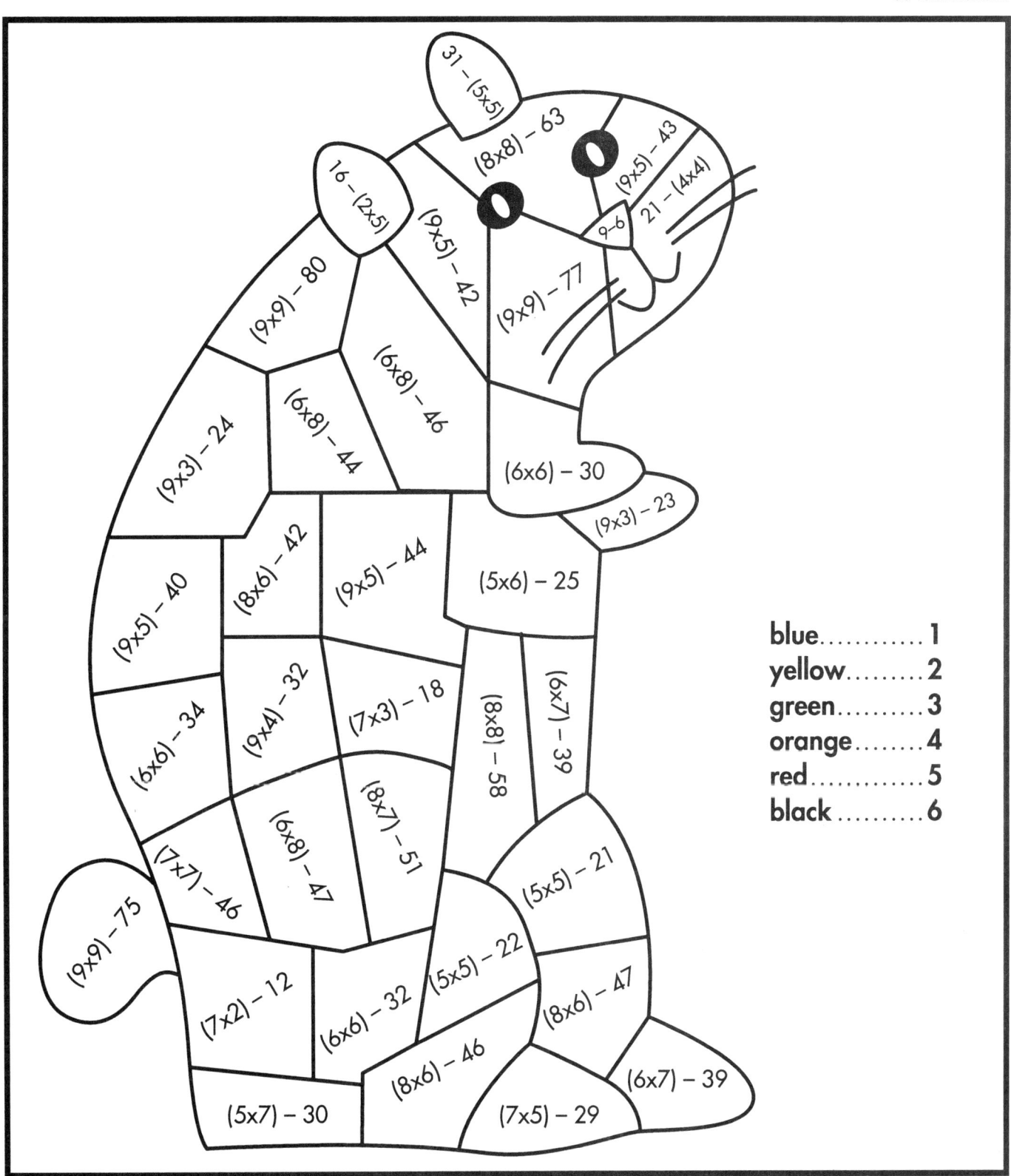

blue............1
yellow..........2
green..........3
orange........4
red............5
black..........6

Multiplication Puzzles ✦ Activity 20

Correction Key

Hamburger

1. Allow your students to correct their own work.
2. Make a transparency of this puzzle and instruct your students to place the transparency over their completed puzzle for a quick and easy check.

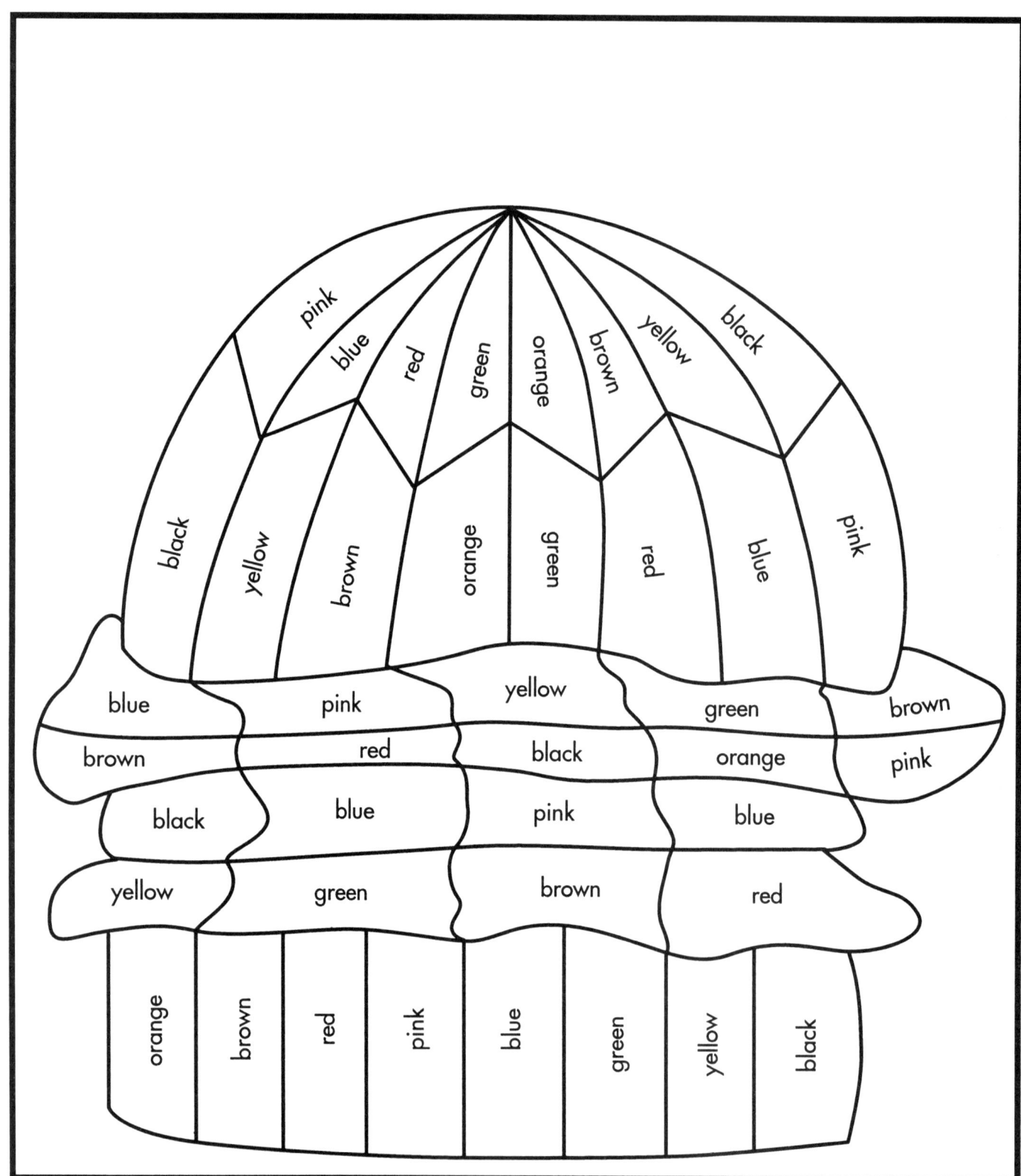

Multiplication Puzzles ✦ Activity 21

42

© Golden Educational Center

Hamburger

Name _____

Date _____

1. Complete the problems within the parentheses first. Then complete the subtraction problem using your answer. (You can do the problems on another piece of paper.)
2. Using your final answer and the color key, color your puzzle correctly.

42 Total Problems

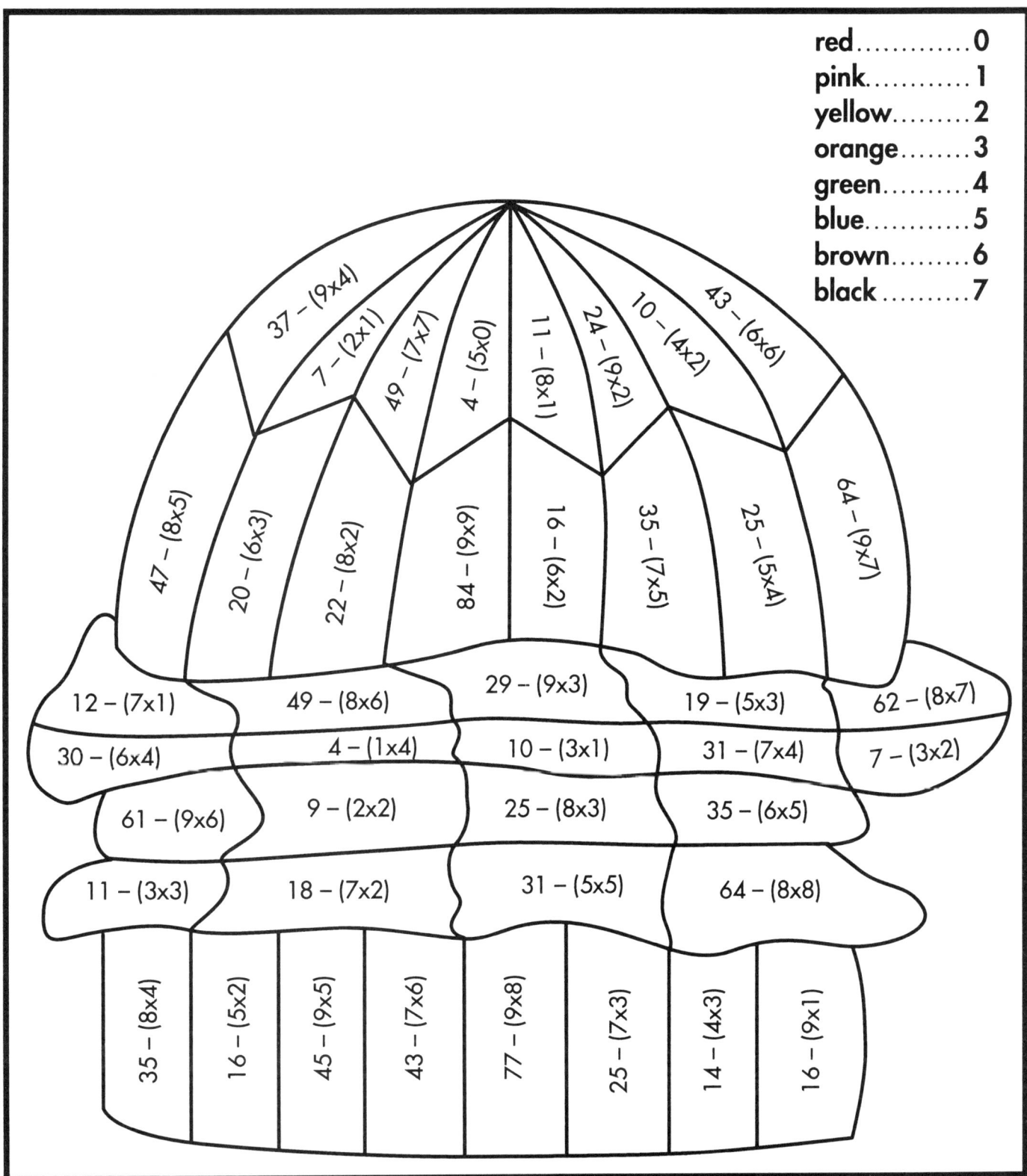

red............0
pink...........1
yellow........2
orange.......3
green.........4
blue...........5
brown........6
black.........7

Multiplication Puzzles ♦ Activity 21

Correction Key

Headdress

1. Allow your students to correct their own work.
2. Make a transparency of this puzzle and instruct your students to place the transparency over their completed puzzle for a quick and easy check.

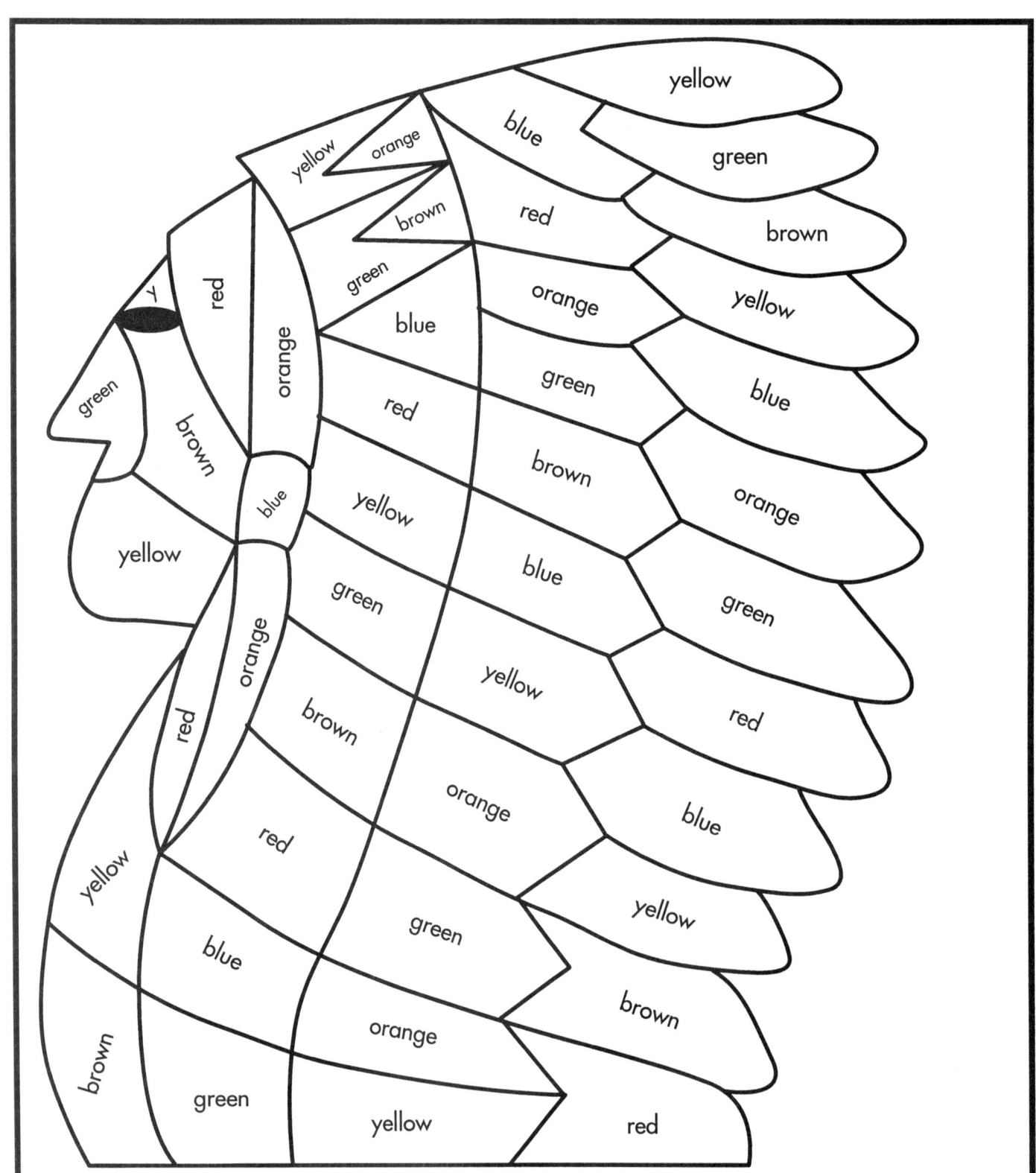

Multiplication Puzzles ✦ Activity 22

Headdress

Name _____

Date _____

1. Complete the problems within the parentheses first. Then complete the or subtraction problem using your answer. (You can do the problems on another piece of paper.)

2. Using your final answer and the color key, color your puzzle correctly.

46 Total Problems

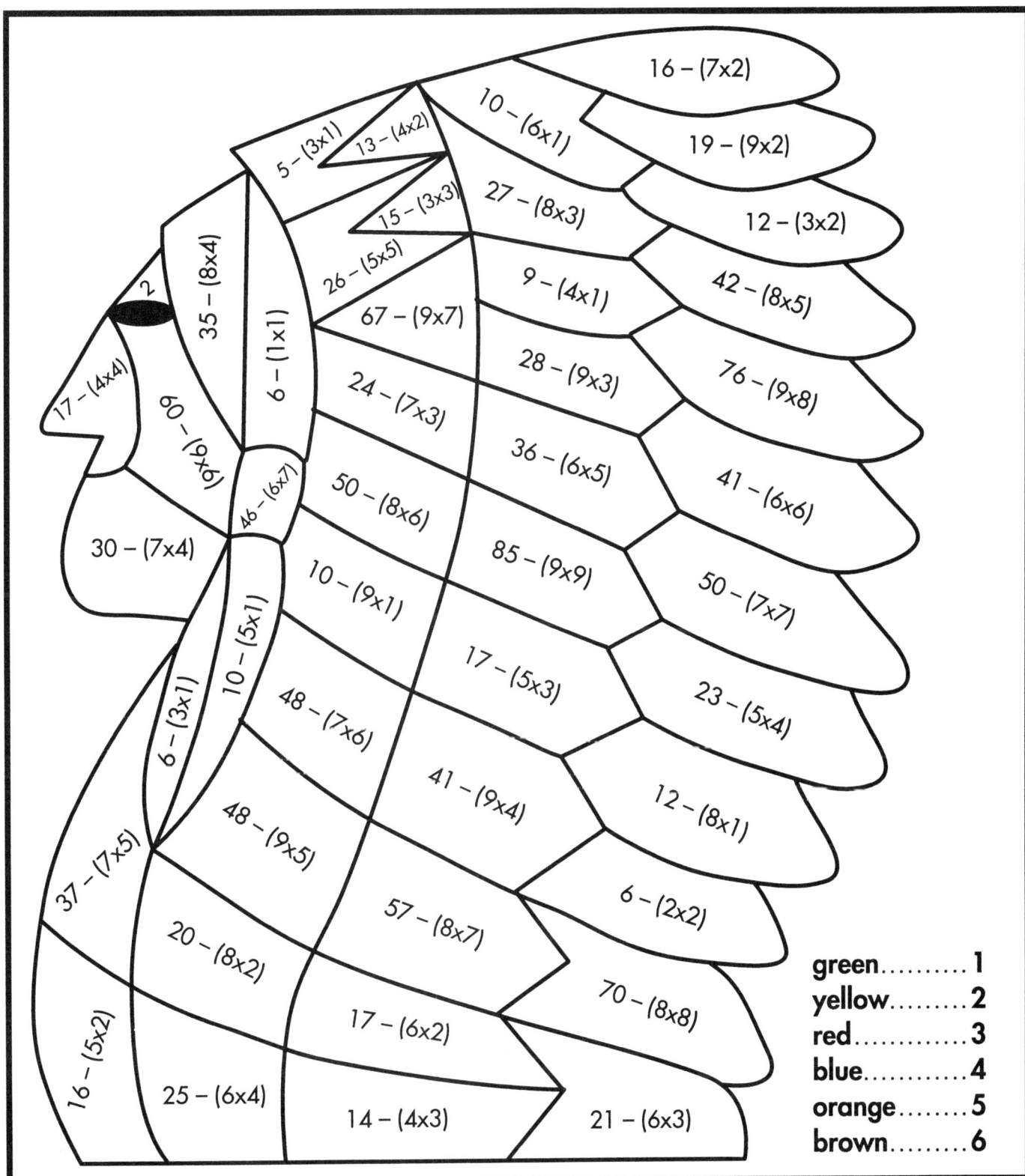

green..........1
yellow.........2
red............3
blue...........4
orange........5
brown.........6

© Golden Educational Center

Multiplication Puzzles ♦ Activity 22

Correction Key

Horse

1. Allow your students to correct their own work.
2. Make a transparency of this puzzle and instruct your students to place the transparency over their completed puzzle for a quick and easy check.

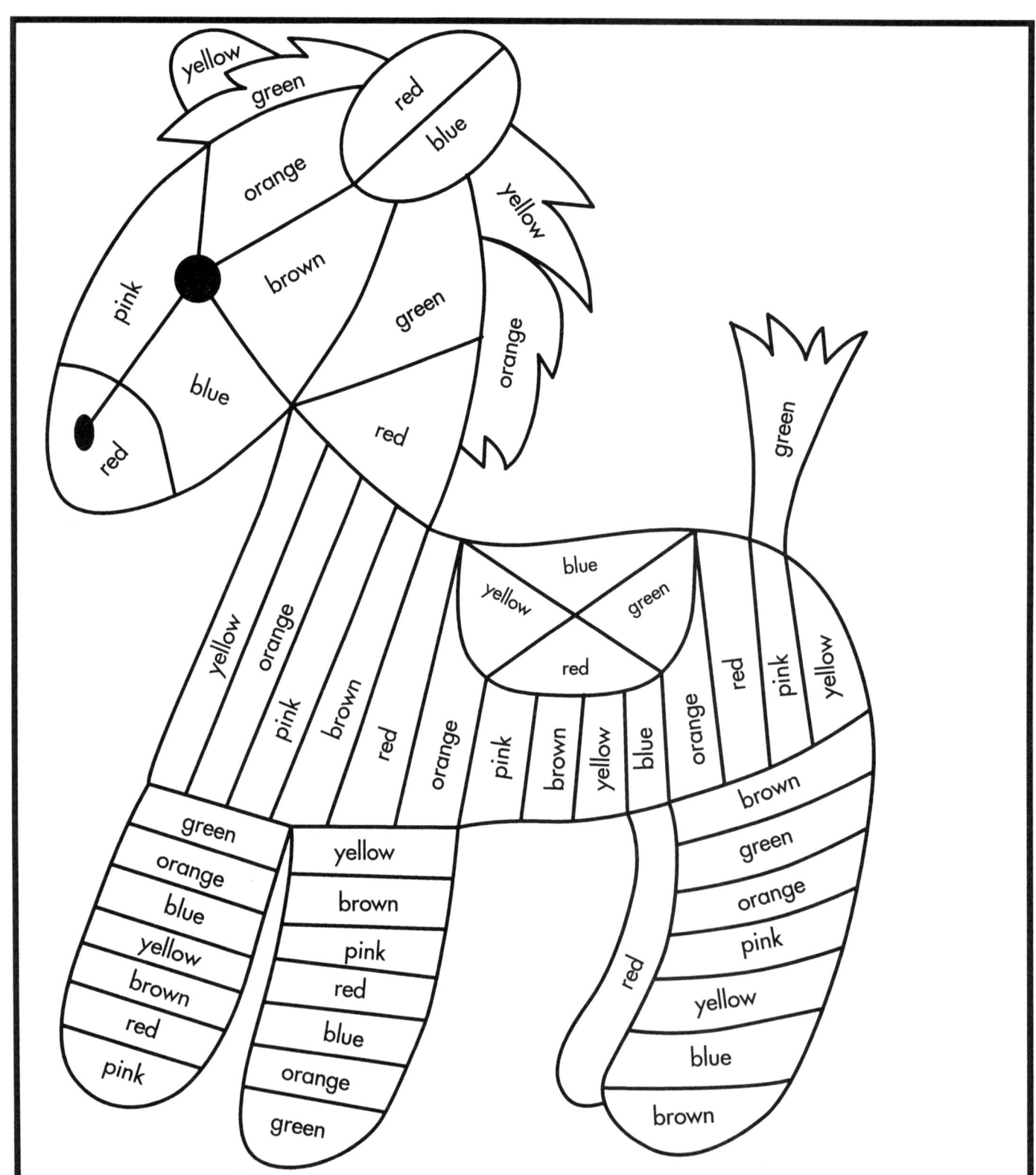

Multiplication Puzzles ✦ Activity 23

Horse

Name _____

Date _____

1. Complete the problems within the parentheses first. Then complete the subtraction problem using your answer. (You can do the problems on another piece of paper.)
2. Using your final answer and the color key, color your puzzle correctly.

53 Total Problems

red............0
blue...........1
green.........2
yellow........3
pink...........4
orange.......5
brown........6

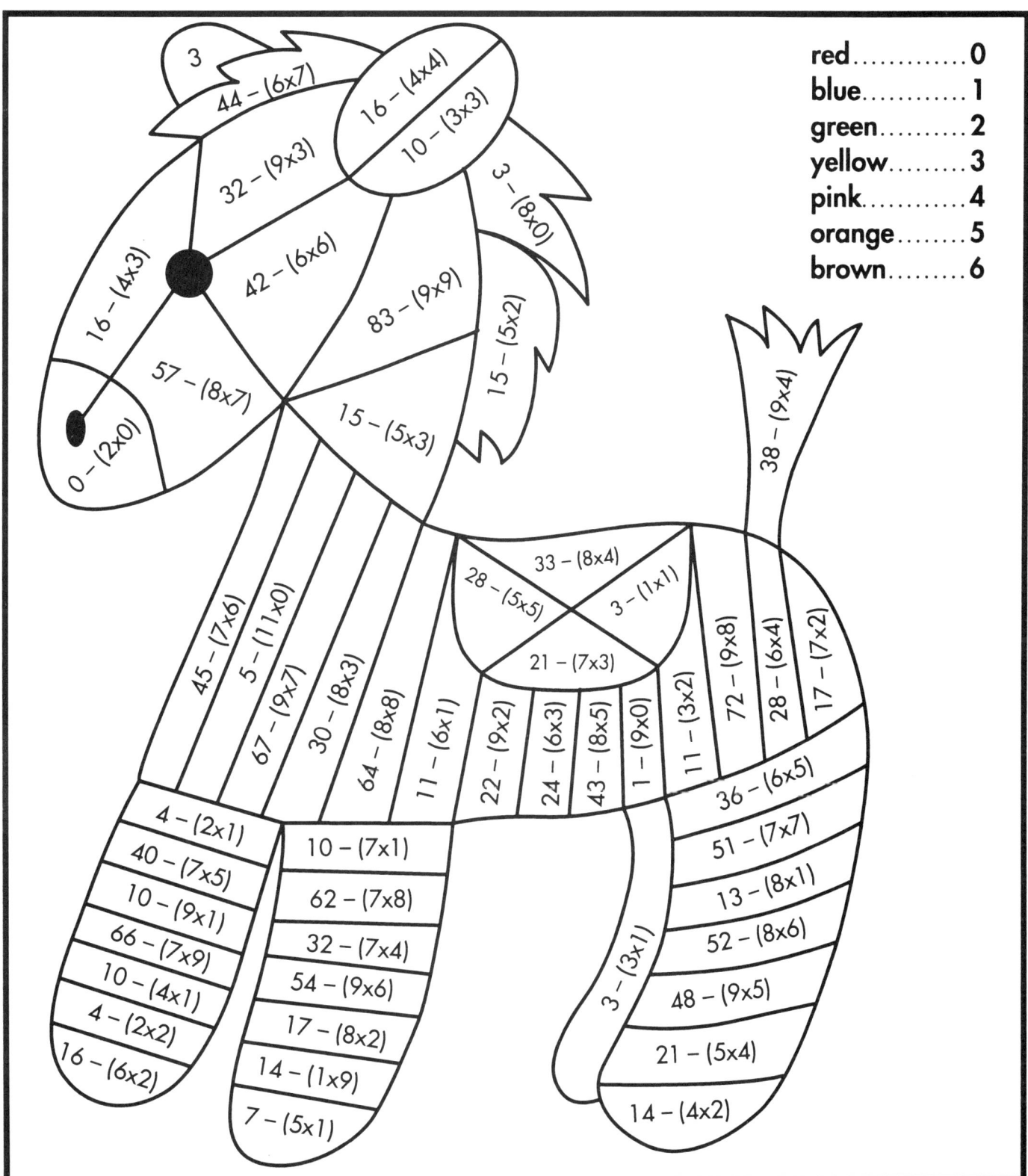

© Golden Educational Center

Multiplication Puzzles ✦ Activity 23

Correction Key

Ice Cream Cone

1. Allow your students to correct their own work.
2. Make a transparency of this puzzle and instruct your students to place the transparency over their completed puzzle for a quick and easy check.

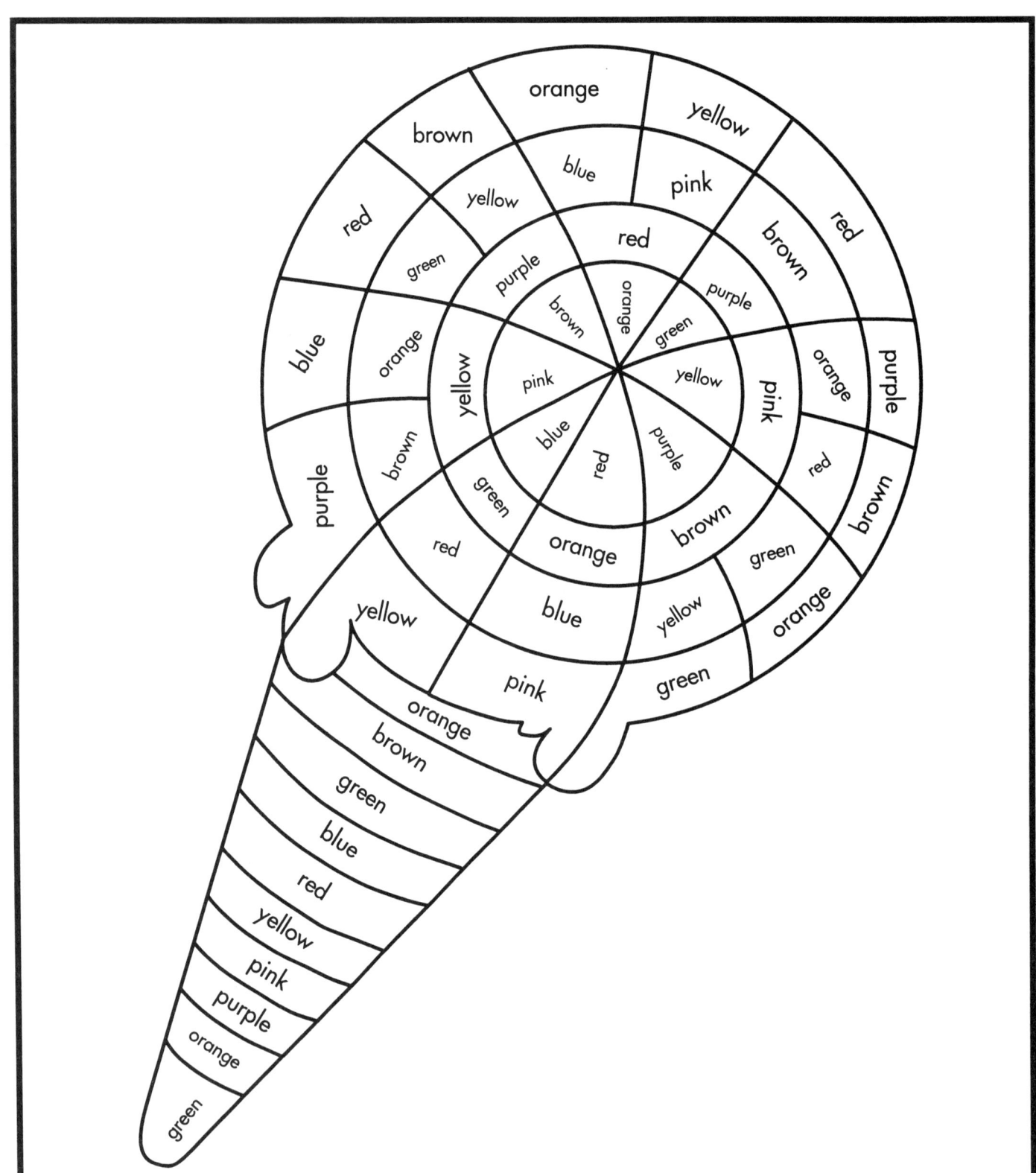

Multiplication Puzzles ✦ **Activity 24**

© Golden Educational Center

Ice Cream Cone

Name _____

Date _____

1. Complete the problems within the parentheses first. Then complete the subtraction problem using your answer. (You can do the problems on another piece of paper.)

2. Using your final answer and the color key, color your puzzle correctly.

51 Total Problems

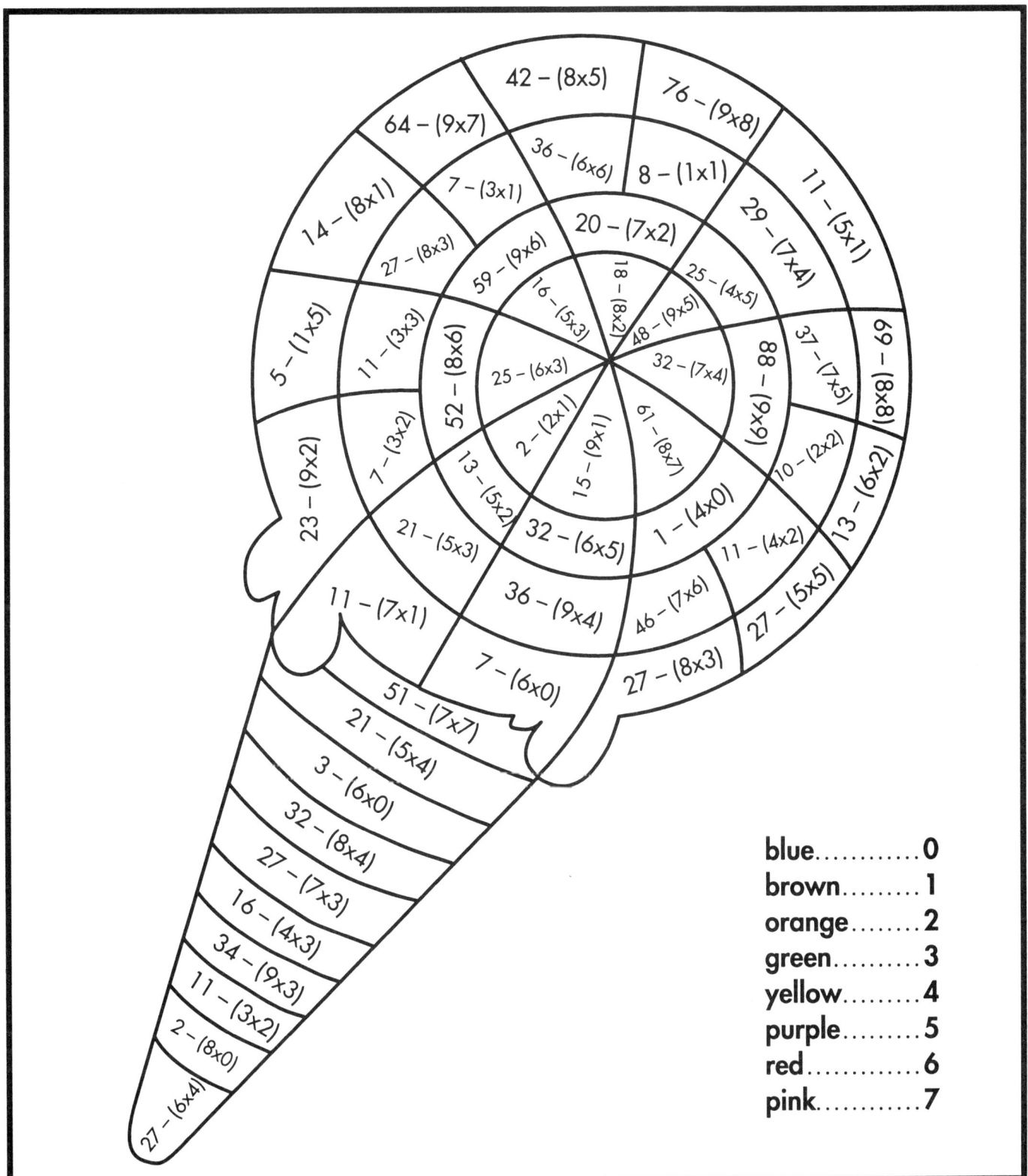

blue............0
brown.........1
orange........2
green..........3
yellow.........4
purple.........5
red.............6
pink............7

© Golden Educational Center

Multiplication Puzzles ✦ Activity 24

Correction Key

July 4th

1. Allow your students to correct their own work.
2. Make a transparency of this puzzle and instruct your students to place the transparency over their completed puzzle for a quick and easy check.

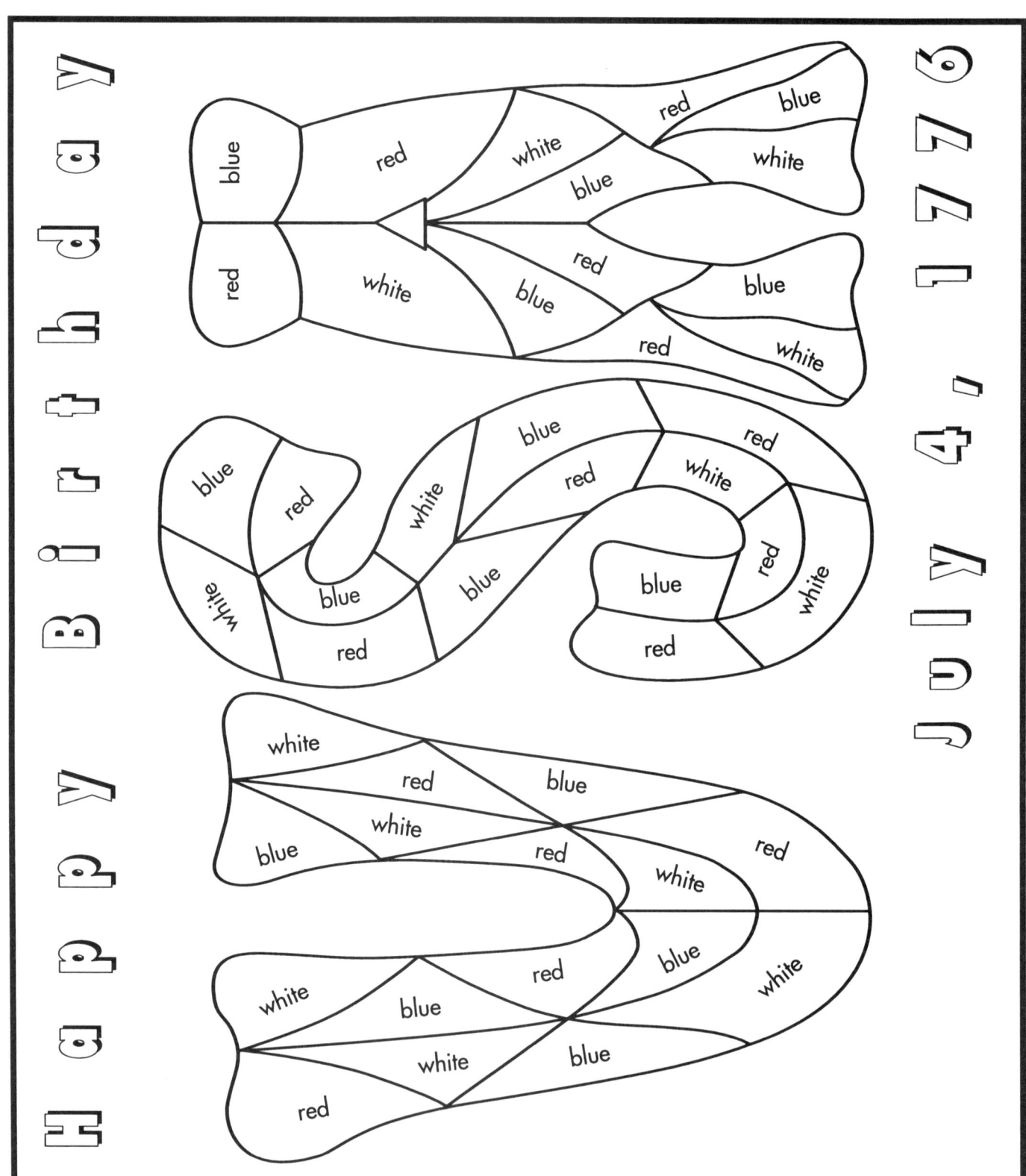

Multiplication Puzzles ✦ Activity 25

© Golden Educational Center

July 4th

Name _____

Date _____

1. Complete the problems within the parentheses first. Then complete the addition or subtraction problem using your answer. (You can do the problems on another piece of paper.)

2. Using your final answer and the color key, color your puzzle correctly.

45 Total Problems

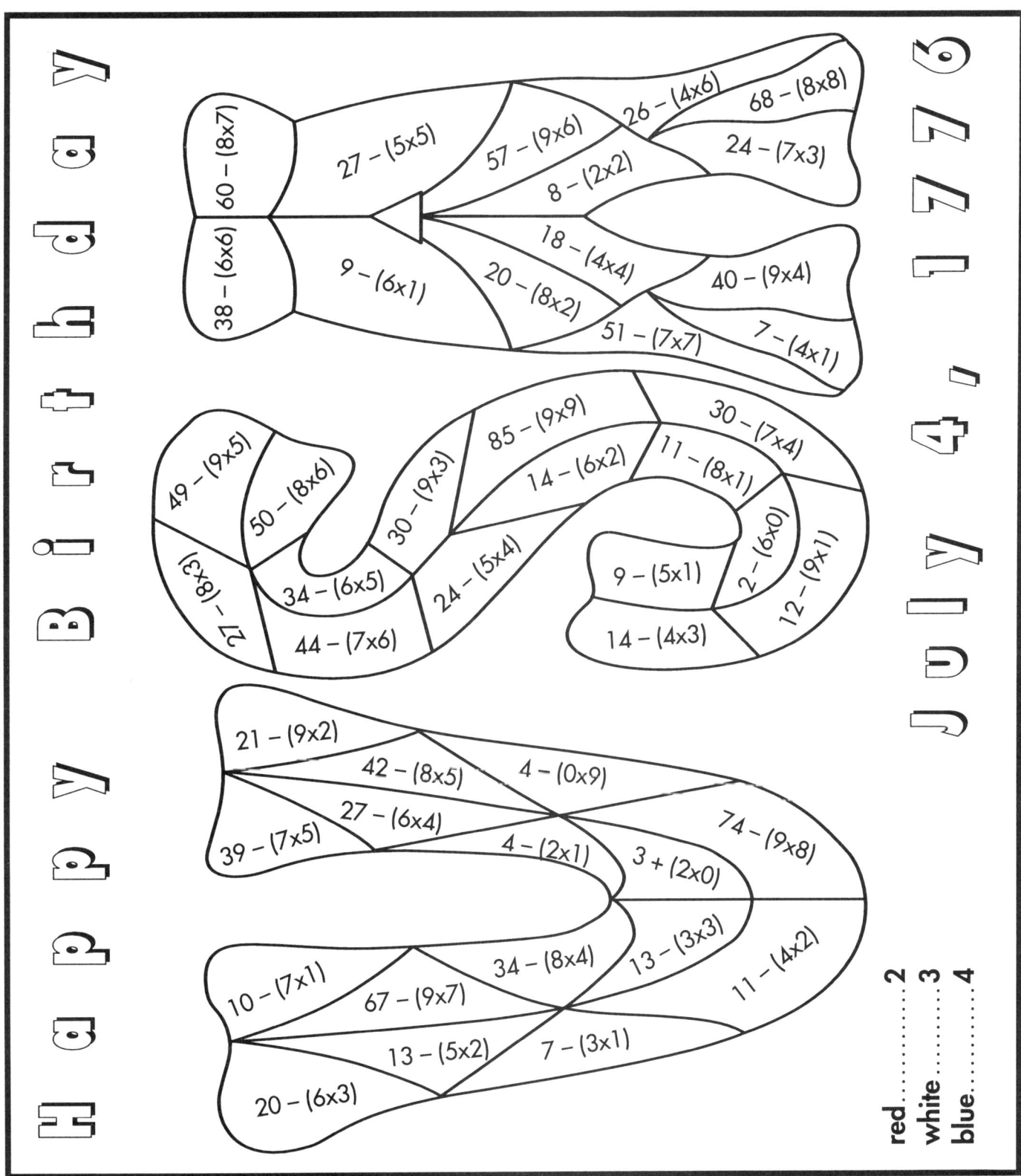

red......2
white....3
blue.....4

Multiplication Puzzles ✦ Activity 25

Correction Key

Kite

1. Allow your students to correct their own work.
2. Make a transparency of this puzzle and instruct your students to place the transparency over their completed puzzle for a quick and easy check.

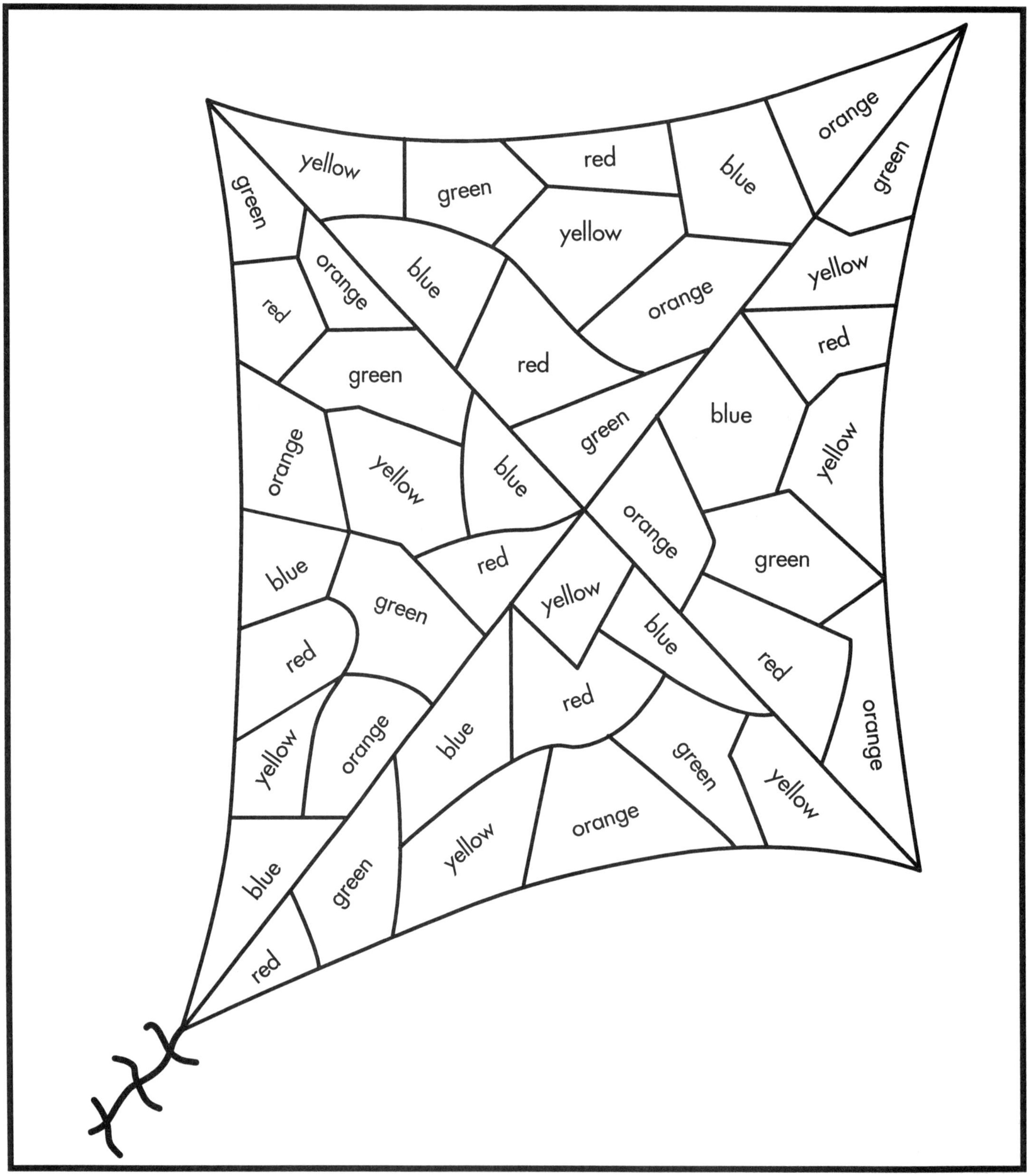

Multiplication Puzzles ✦ Activity 26

Kite

Name _____

Date _____

1. Complete the problems within the parentheses first. Then complete the subtraction problem using your answer. (You can do the problems on another piece of paper.)

2. Using your final answer and the color key, color your puzzle correctly.

43 Total Problems

Puzzle regions (subtraction problems):
- 13 – (3x2)
- 32 – (9x3)
- 19 – (4x3)
- 43 – (8x5)
- 69 – (9x7)
- 40 – (9x4)
- 16 – (3x3)
- 21 – (9x2)
- 4 – (9x0)
- 42 – (6x6)
- 40 – (7x5)
- 49 – (9x5)
- 21 – (4x4)
- 35 – (7x4)
- 75 – (9x8)
- 15 – (6x2)
- 34 – (6x5)
- 37 – (8x4)
- 62 – (8x7)
- 25 – (6x3)
- 54 – (8x6)
- 47 – (7x6)
- 60 – (9x6)
- 22 – (5x3)
- 10 – (7x1)
- 9 – (4x1)
- 24 – (5x4)
- 30 – (6x4)
- 21 – (7x2)
- 52 – (7x7)
- 24 – (7x3)
- 8 – (3x1)
- 8 – (2x2)
- 22 – (8x2)
- 67 – (8x8)
- 31 – (8x3)
- 86 – (9x9)
- 12 – (8x1)
- 7 – (1x1)
- 3 – (6x0)
- 15 – (4x2)
- 15 – (5x2)
- 29 – (5x5)

Color key:

red............3
orange........4
yellow.........5
blue............6
green..........7

Multiplication Puzzles ✦ Activity 26

Correction Key

Kitten

1. Allow your students to correct their own work.
2. Make a transparency of this puzzle and instruct your students to place the transparency over their completed puzzle for a quick and easy check.

Multiplication Puzzles ✦ Activity 27

Kitten

Name _____

Date _____

1. Complete the problems within the parentheses first. Then complete the subtraction problem using your answer. (You can do the problems on another piece of paper.)
2. Using your final answer and the color key, color your puzzle correctly.

50 Total Problems

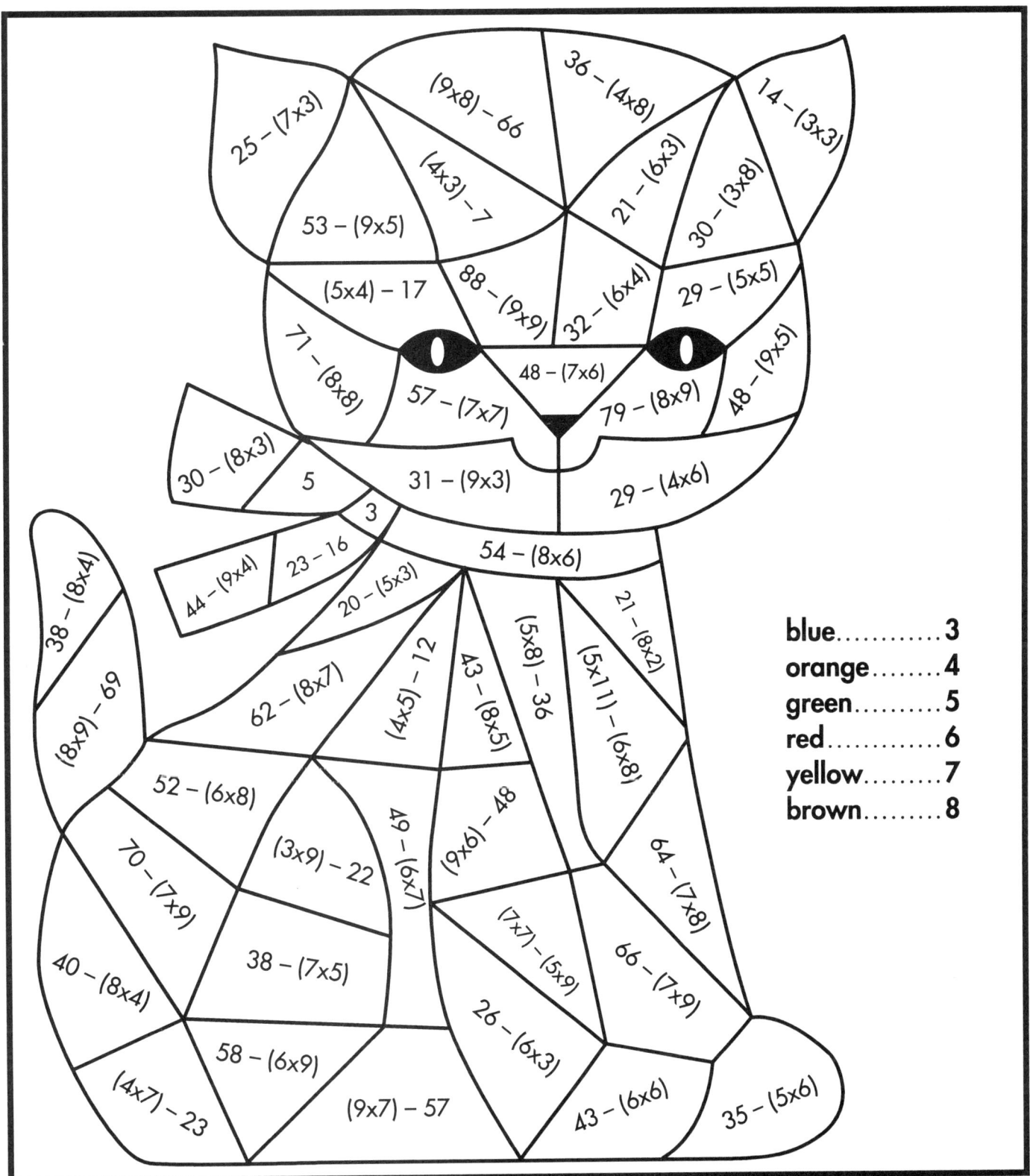

blue..........3
orange........4
green.........5
red...........6
yellow........7
brown.........8

© Golden Educational Center

Multiplication Puzzles ✦ Activity 27

Correction Key   **Ladybug**

1. Allow your students to correct their own work.
2. Make a transparency of this puzzle and instruct your students to place the transparency over their completed puzzle for a quick and easy check.

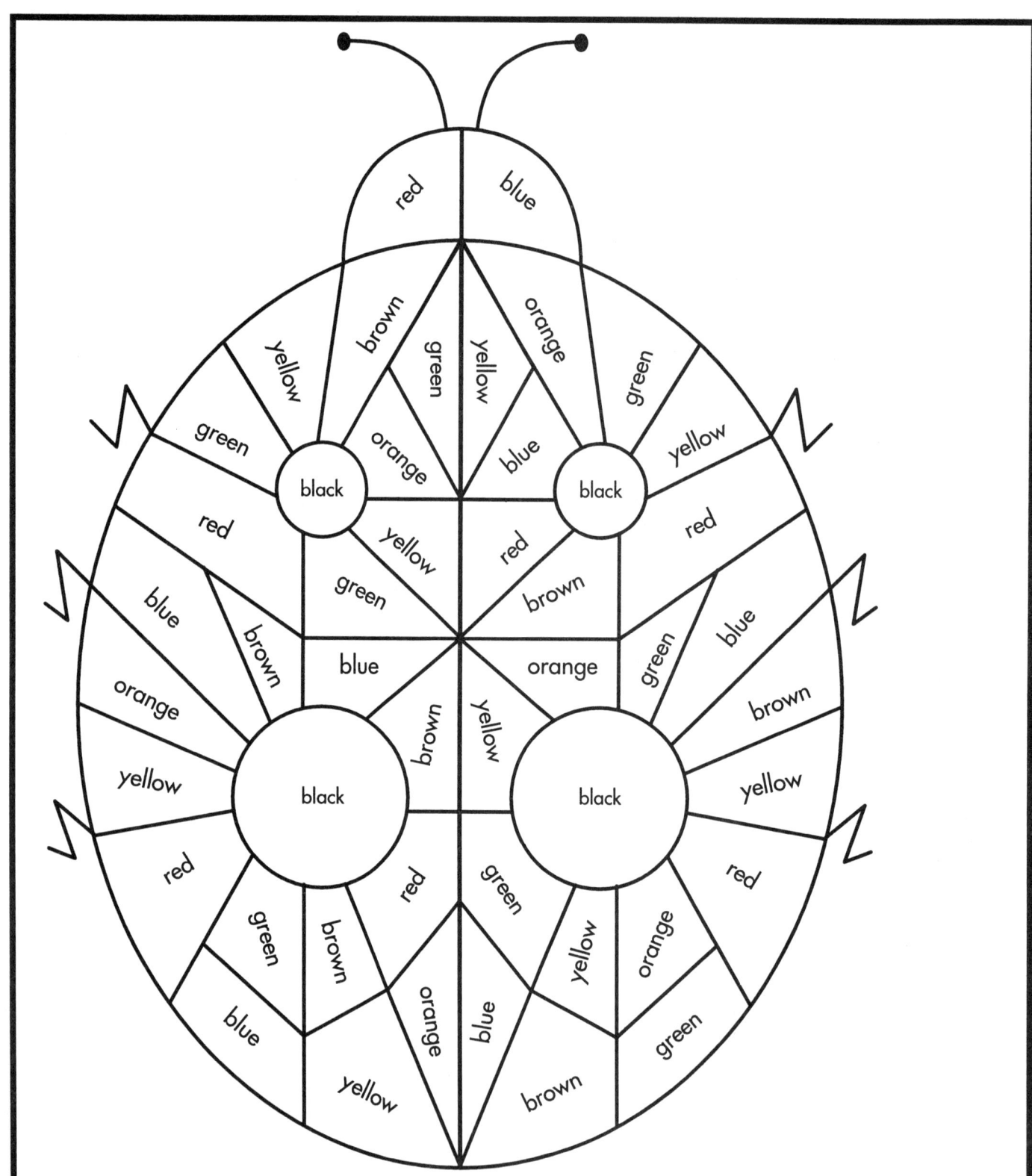

Multiplication Puzzles ✦ Activity 28

Ladybug

Name _____

Date _____

1. Complete the problems within the parentheses first. Then complete the addition or subtraction problem using your answer. (You can do the problems on another piece of paper.)

2. Using your final answer and the color key, color your puzzle correctly.

48 Total Problems

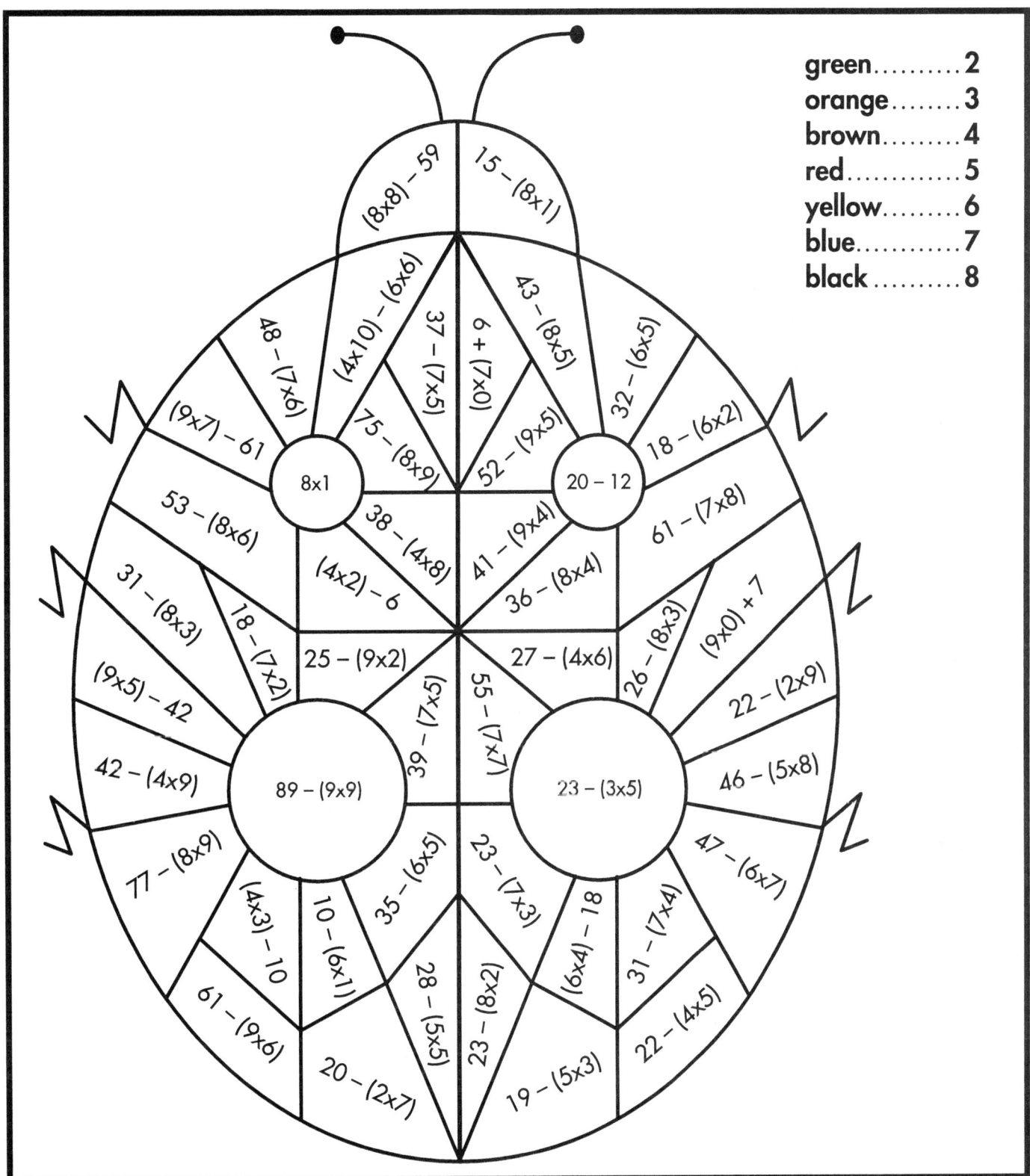

green..........2
orange........3
brown.........4
red.............5
yellow.........6
blue............7
black..........8

Multiplication Puzzles ◆ Activity 28

Correction Key   Leaf

1. Allow your students to correct their own work.
2. Make a transparency of this puzzle and instruct your students to place the transparency over their completed puzzle for a quick and easy check.

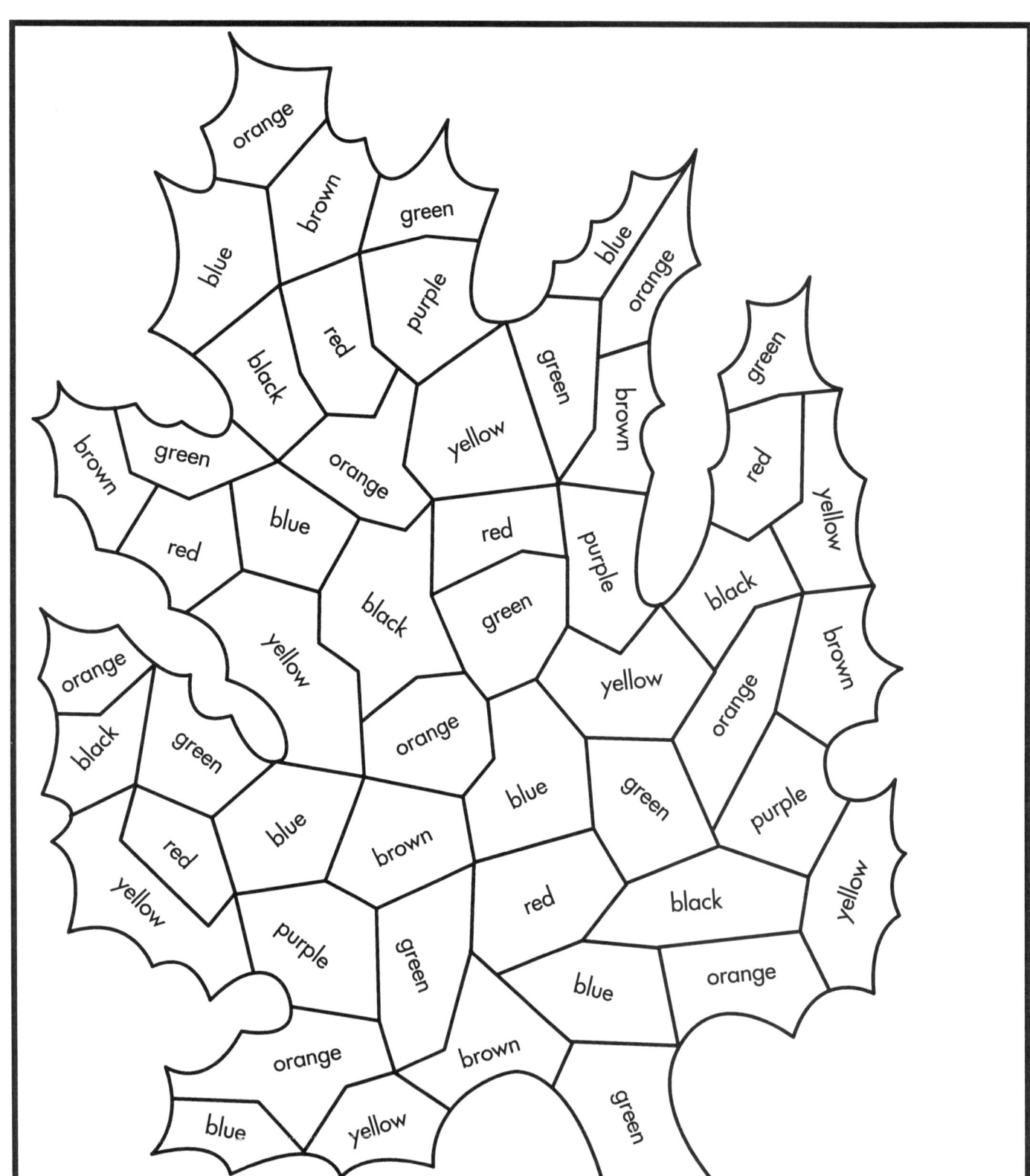

Multiplication Puzzles ♦ Activity 29

Leaf

Name _____

Date _____

1. Complete the problems within the parentheses first. Then complete the subtraction problem using your answer. (You can do the problems on another piece of paper.)
2. Using your final answer and the color key, color your puzzle correctly.

52 Total Problems

blue............1
orange........2
purple.........3
red.............4
yellow.........5
black..........6
green..........7
brown.........8

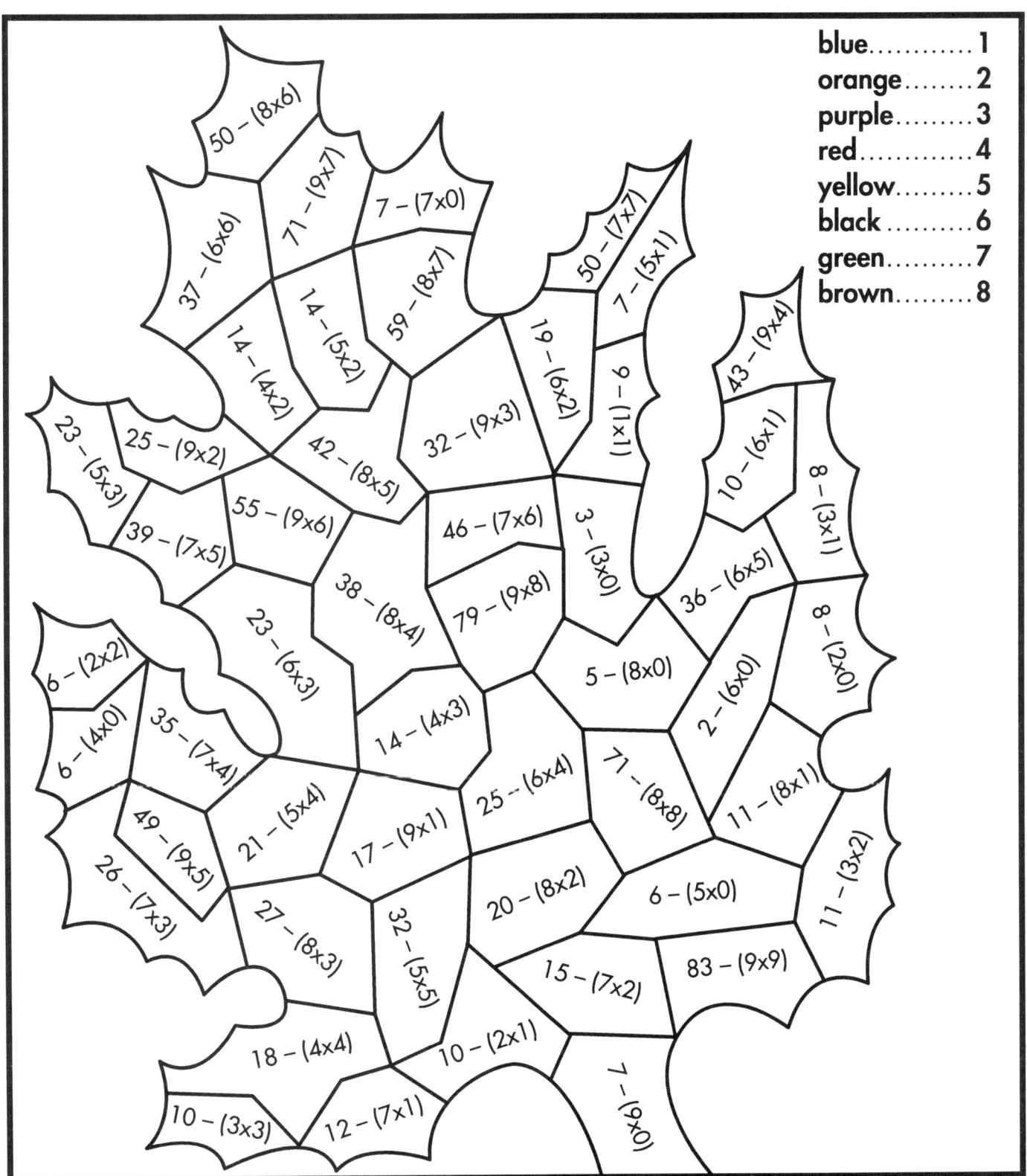

Multiplication Puzzles ✦ Activity 29

Correction Key

Liberty Bell

1. Allow your students to correct their own work.
2. Make a transparency of this puzzle and instruct your students to place the transparency over their completed puzzle for a quick and easy check.

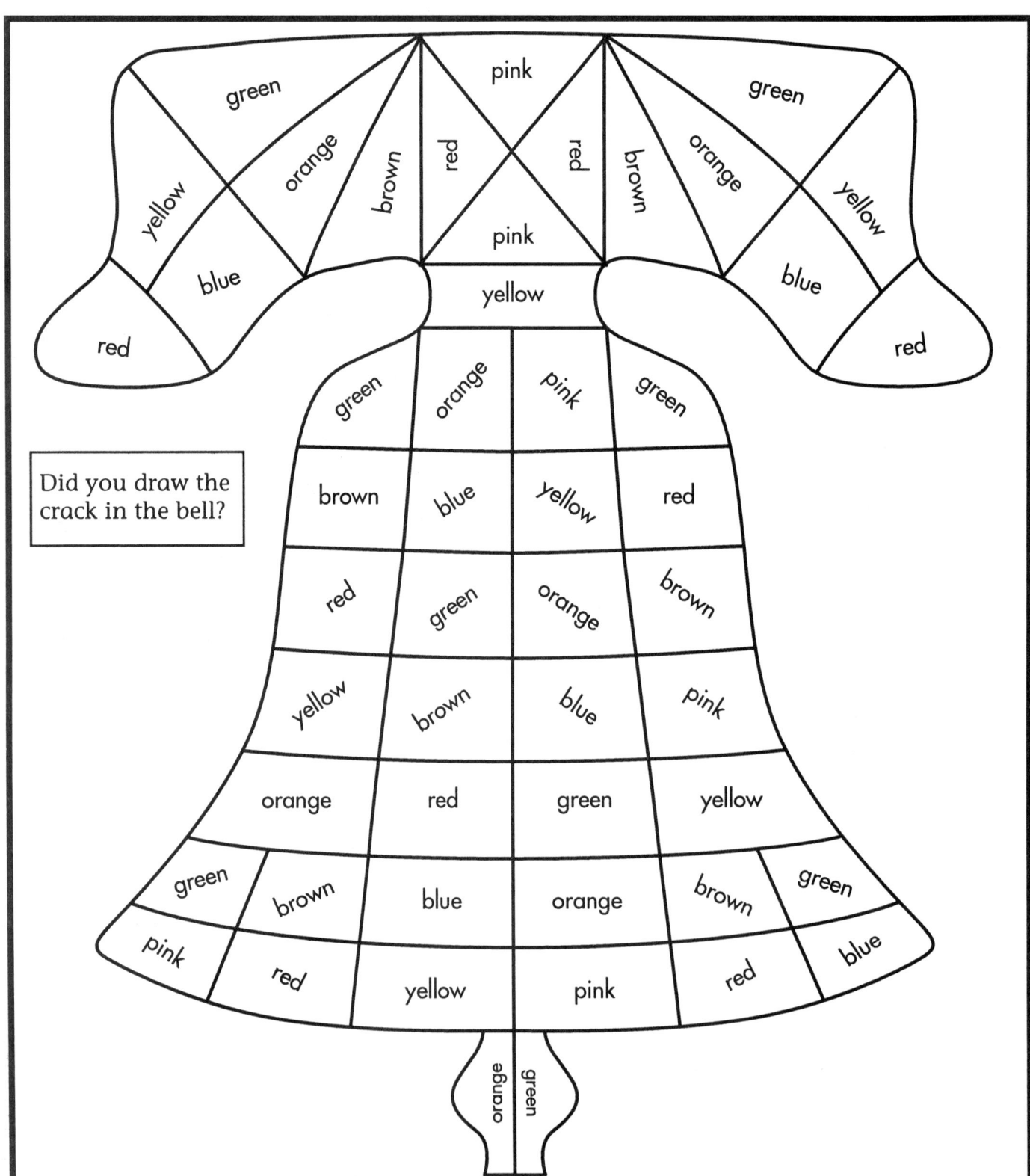

Did you draw the crack in the bell?

Multiplication Puzzles ✦ Activity 30

Liberty Bell

Name _____

Date _____

1. Complete the problems within the parentheses first. Then complete the subtraction problem using your answer. (You can do the problems on another piece of paper.)

2. Using your final answer and the color key, color your puzzle correctly.

51 Total Problems

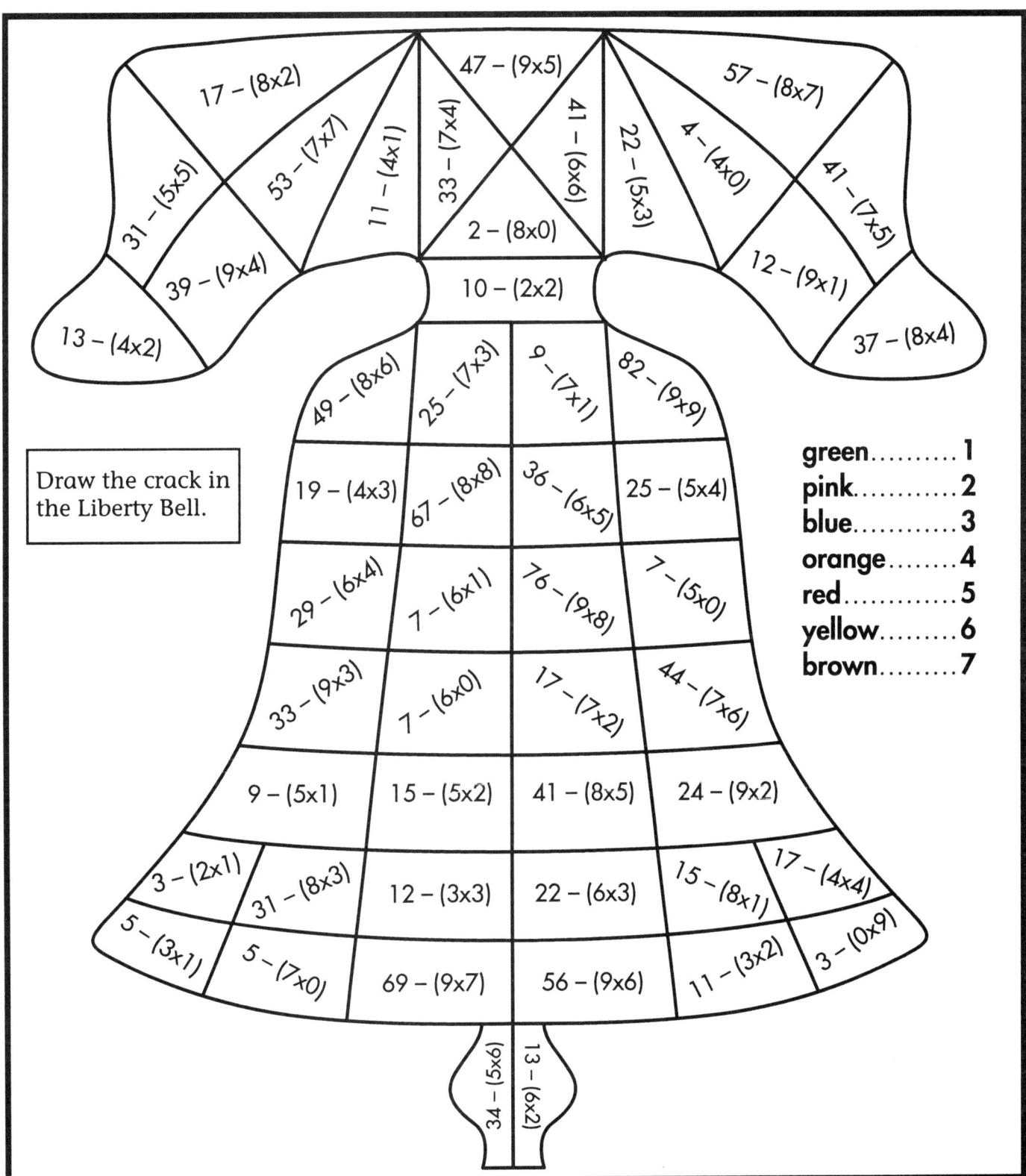

Draw the crack in the Liberty Bell.

green..........1
pink............2
blue............3
orange........4
red.............5
yellow.........6
brown.........7

Multiplication Puzzles ♦ Activity 30

Correction Key

Lily

1. Allow your students to correct their own work.
2. Make a transparency of this puzzle and instruct your students to place the transparency over their completed puzzle for a quick and easy check.

Multiplication Puzzles ✦ Activity 31

Lily

Name _____

Date _____

1. Complete the problems within the parentheses first. Then complete the subtraction problem using your answer. (You can do the problems on another piece of paper.)
2. Using your final answer and the color key, color your puzzle correctly.

53 Total Problems

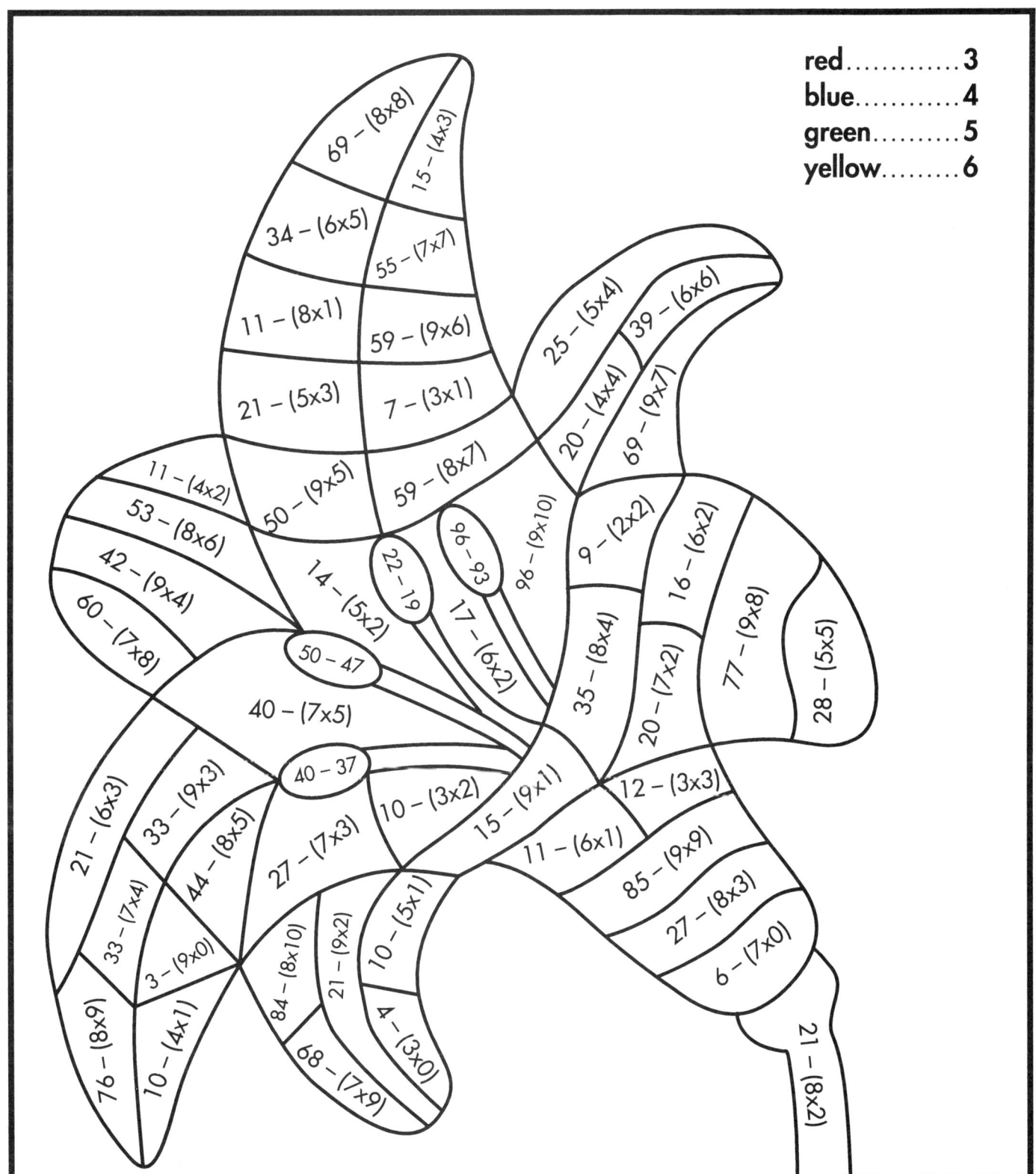

red 3
blue 4
green 5
yellow 6

Multiplication Puzzles ✦ Activity 31

Correction Key

Lion

1. Allow your students to correct their own work.
2. Make a transparency of this puzzle and instruct your students to place the transparency over their completed puzzle for a quick and easy check.

Multiplication Puzzles ✦ Activity 32

Lion

Name _____

Date _____

1. Complete the problems within the parentheses first. Then complete or subtraction problem using your answer. (You can do the problems on another piece of paper.)
2. Using your final answer and the color key, color your puzzle correctly.

47 Total Problems

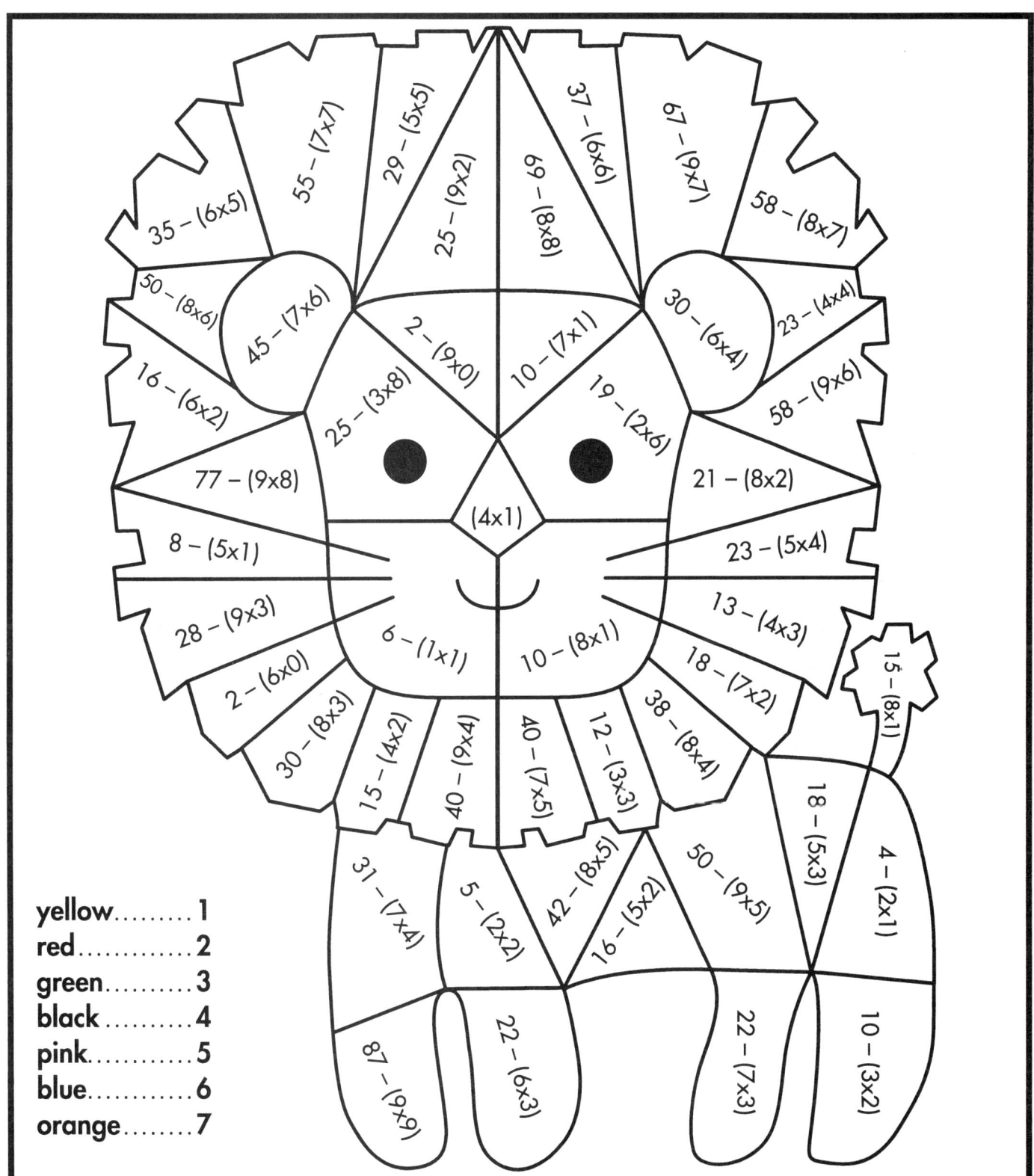

yellow 1
red 2
green 3
black 4
pink 5
blue 6
orange 7

Multiplication Puzzles ✦ Activity 32

Correction Key

Magnolia Flower

1. Allow your students to correct their own work.
2. Make a transparency of this puzzle and instruct your students to place the transparency over their completed puzzle for a quick and easy check.

Multiplication Puzzles ✦ Activity 33

Magnolia Flower

Name _____

Date _____

1. Complete the problems within the parentheses first. Then complete the subtraction problem using your answer. (You can do the problems on another piece of paper.)
2. Using your final answer and the color key, color your puzzle correctly.

46 Total Problems

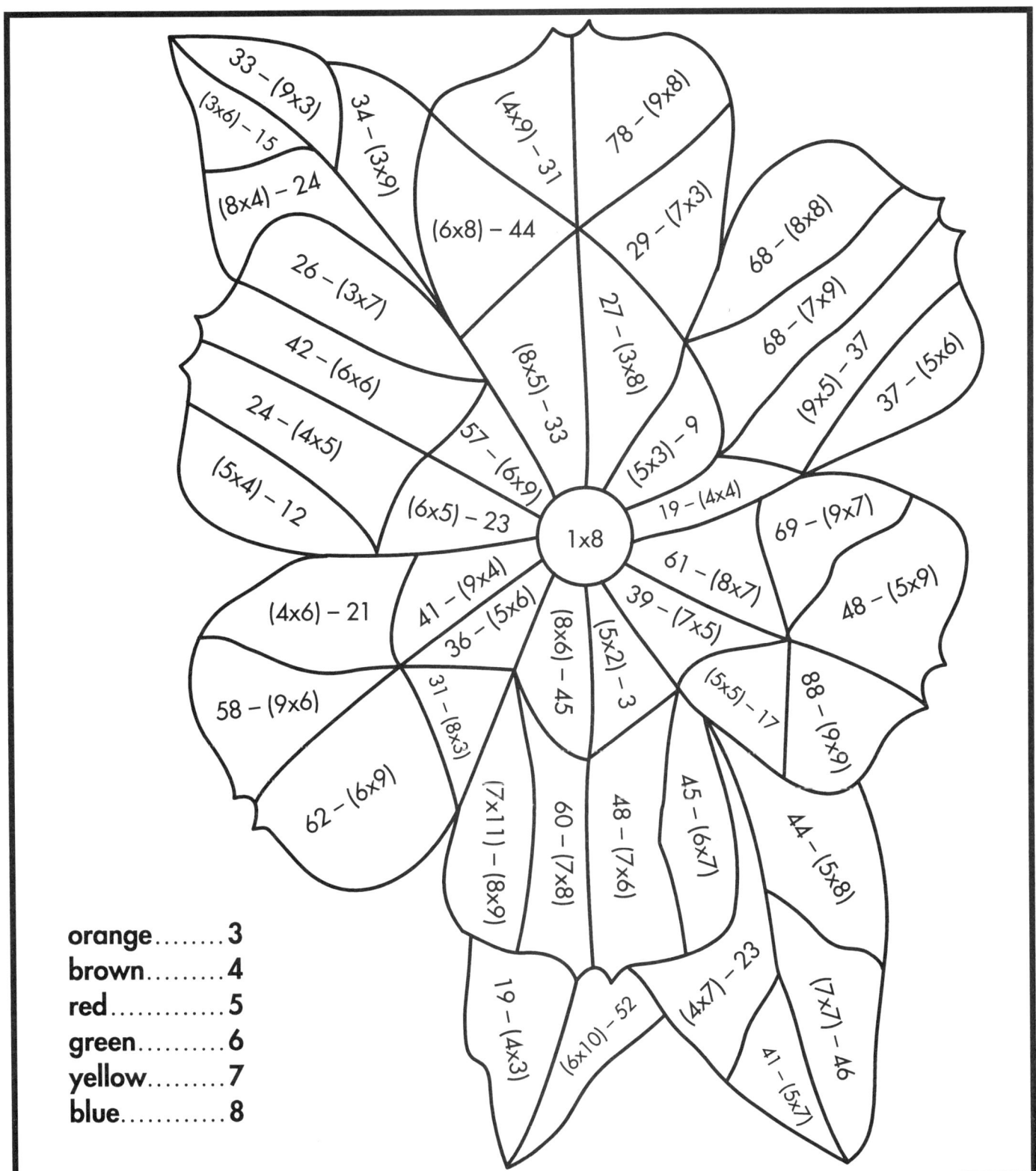

orange....... 3
brown........ 4
red............. 5
green.......... 6
yellow........ 7
blue............ 8

Multiplication Puzzles ✦ Activity 33

Correction Key

Moon

1. Allow your students to correct their own work.
2. Make a transparency of this puzzle and instruct your students to place the transparency over their completed puzzle for a quick and easy check.

Multiplication Puzzles ✦ Activity 34

68

© Golden Educational Center

Moon

Name _____

Date _____

1. Complete the problems within the parentheses first. Then complete the subtraction problem using your answer. (You can do the problems on another piece of paper.)
2. Using your final answer and the color key, color your puzzle correctly.

60 Total Problems

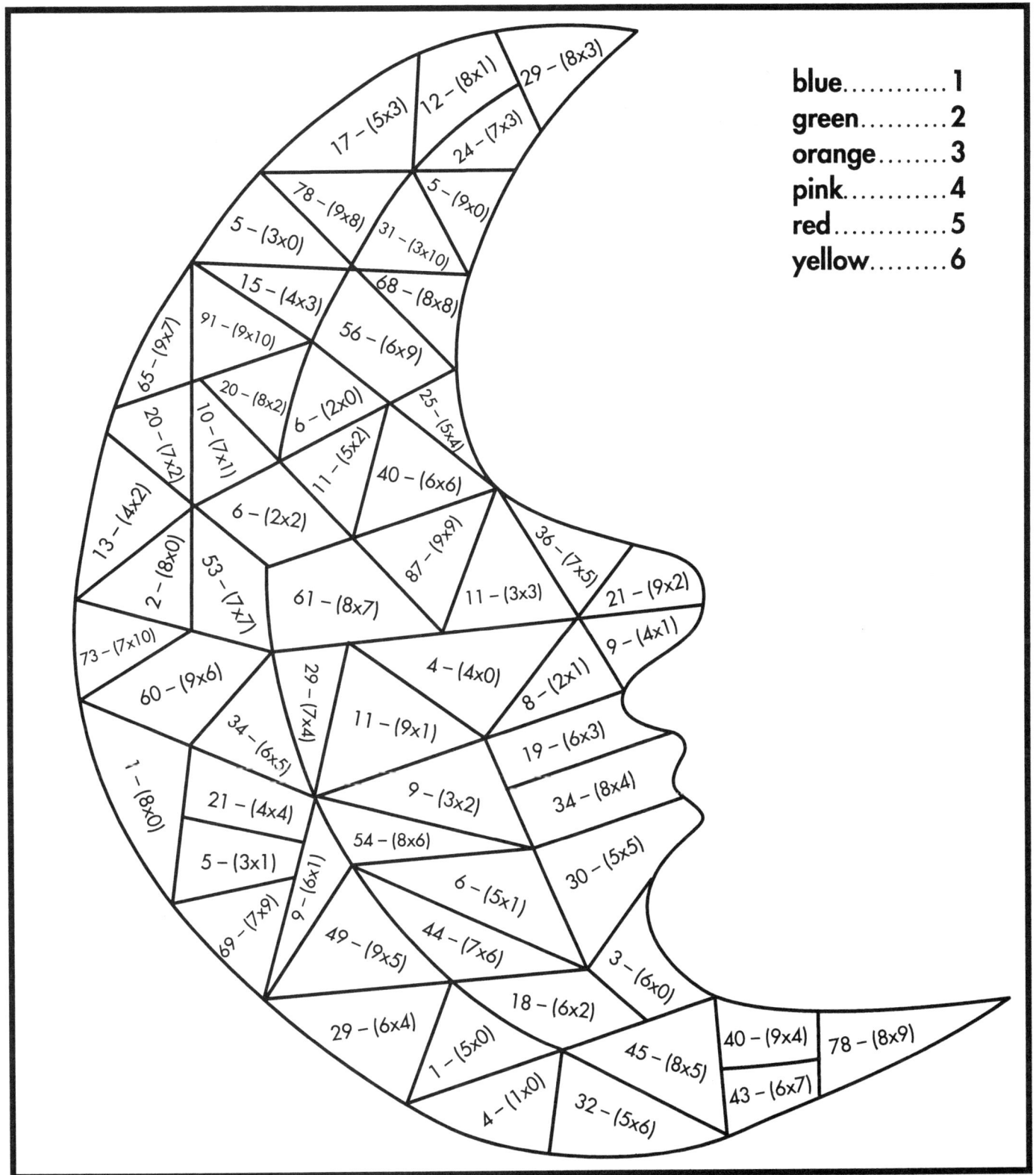

blue............1
green..........2
orange........3
pink............4
red..............5
yellow........6

© GOLDEN EDUCATIONAL CENTER

Multiplication Puzzles ♦ Activity 34

Correction Key

Mouse

1. Allow your students to correct their own work.
2. Make a transparency of this puzzle and instruct your students to place the transparency over their completed puzzle for a quick and easy check.

Multiplication Puzzles ✦ Activity 35

Mouse

Name _____

Date _____

1. Complete the problems within the parentheses first. Then complete the subtraction problem using your answer. (You can do the problems on another piece of paper.)

2. Using your final answer and the color key, color your puzzle correctly.

49 Total Problems

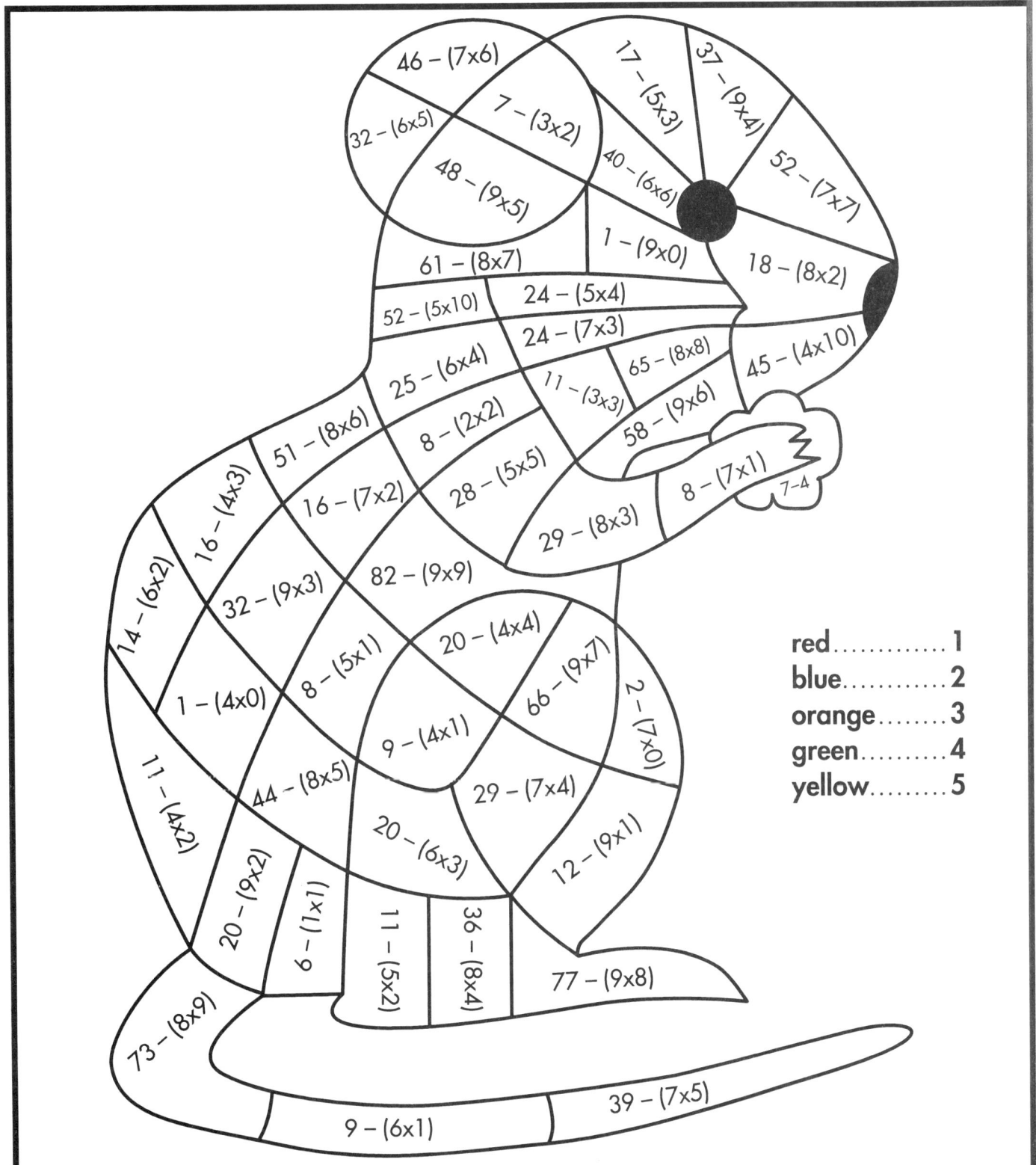

red 1
blue 2
orange 3
green 4
yellow 5

Multiplication Puzzles ✦ Activity 35

Correction Key

Mushrooms

1. Allow your students to correct their own work.
2. Make a transparency of this puzzle and instruct your students to place the transparency over their completed puzzle for a quick and easy check.

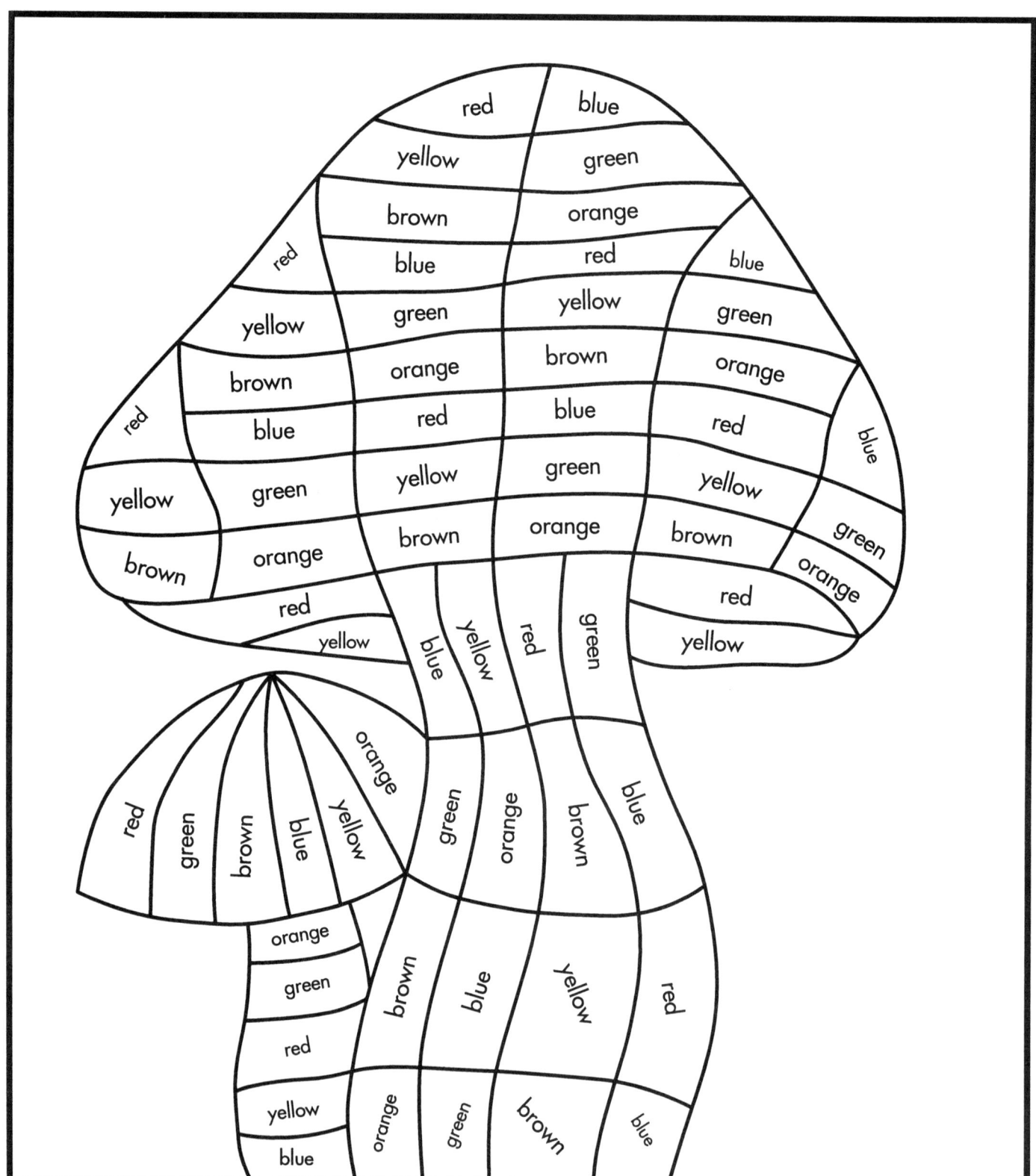

Multiplication Puzzles ✦ Activity 36

Mushrooms

Name _____

Date _____

1. Complete the problems within the parentheses first. Then complete the addition or subtraction problem using your answer. (You can do the problems on another piece of paper.)

2. Using your final answer and the color key, color your puzzle correctly.

67 Total Problems

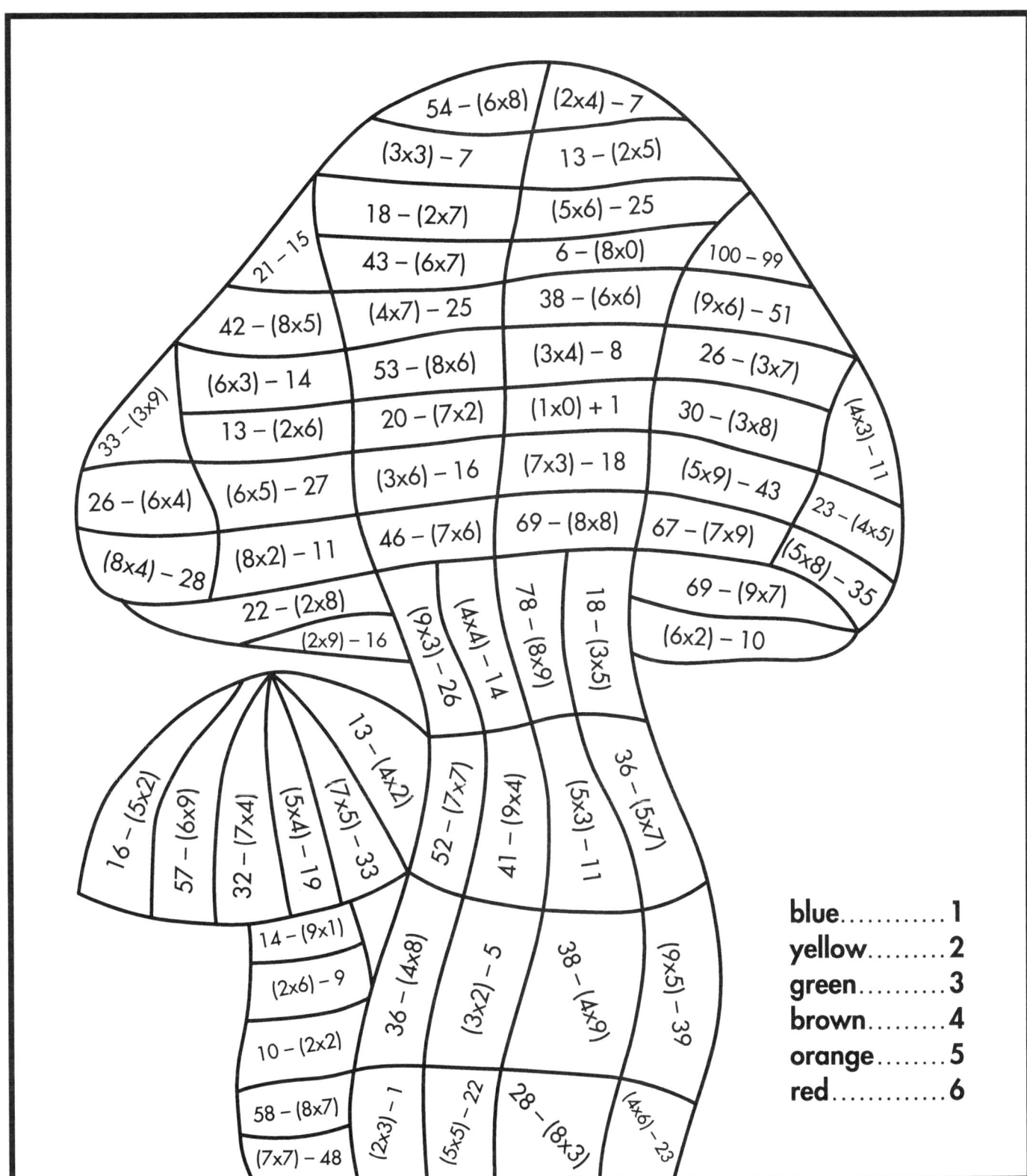

blue............1
yellow..........2
green...........3
brown..........4
orange.........5
red.............6

Multiplication Puzzles ♦ Activity 36

Correction Key

Numbers

1. Allow your students to correct their own work.
2. Make a transparency of this puzzle and instruct your students to place the transparency over their completed puzzle for a quick and easy check.

Multiplication Puzzles ✦ Activity 37

Numbers

Name _____

Date _____

1. Complete the problems within the parentheses first. Then complete the subtraction problem using your answer. (You can do the problems on another piece of paper.)

2. Using your final answer and the color key, color your puzzle correctly.

58 Total Problems

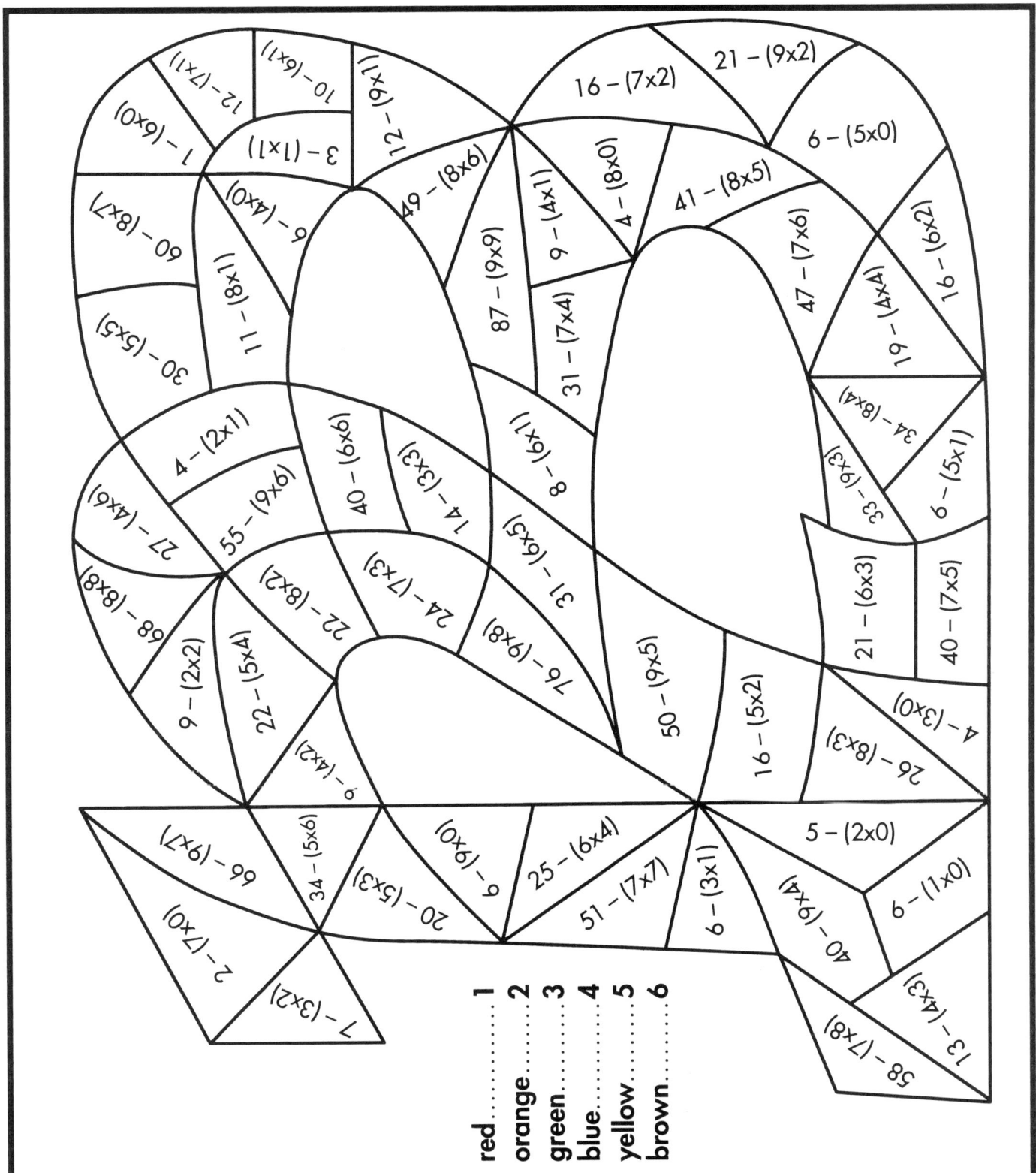

Multiplication Puzzles ✦ Activity 37

Correction Key

Penguin

1. Allow your students to correct their own work.
2. Make a transparency of this puzzle and instruct your students to place the transparency over their completed puzzle for a quick and easy check.

Multiplication Puzzles ✦ Activity 38

Penguin

Name _____

Date _____

1. Complete the problems within the parentheses first. Then complete the addition or subtraction problem using your answer. (You can do the problems on another piece of paper.)
2. Using your final answer and the color key, color your puzzle correctly.

31 Total Problems

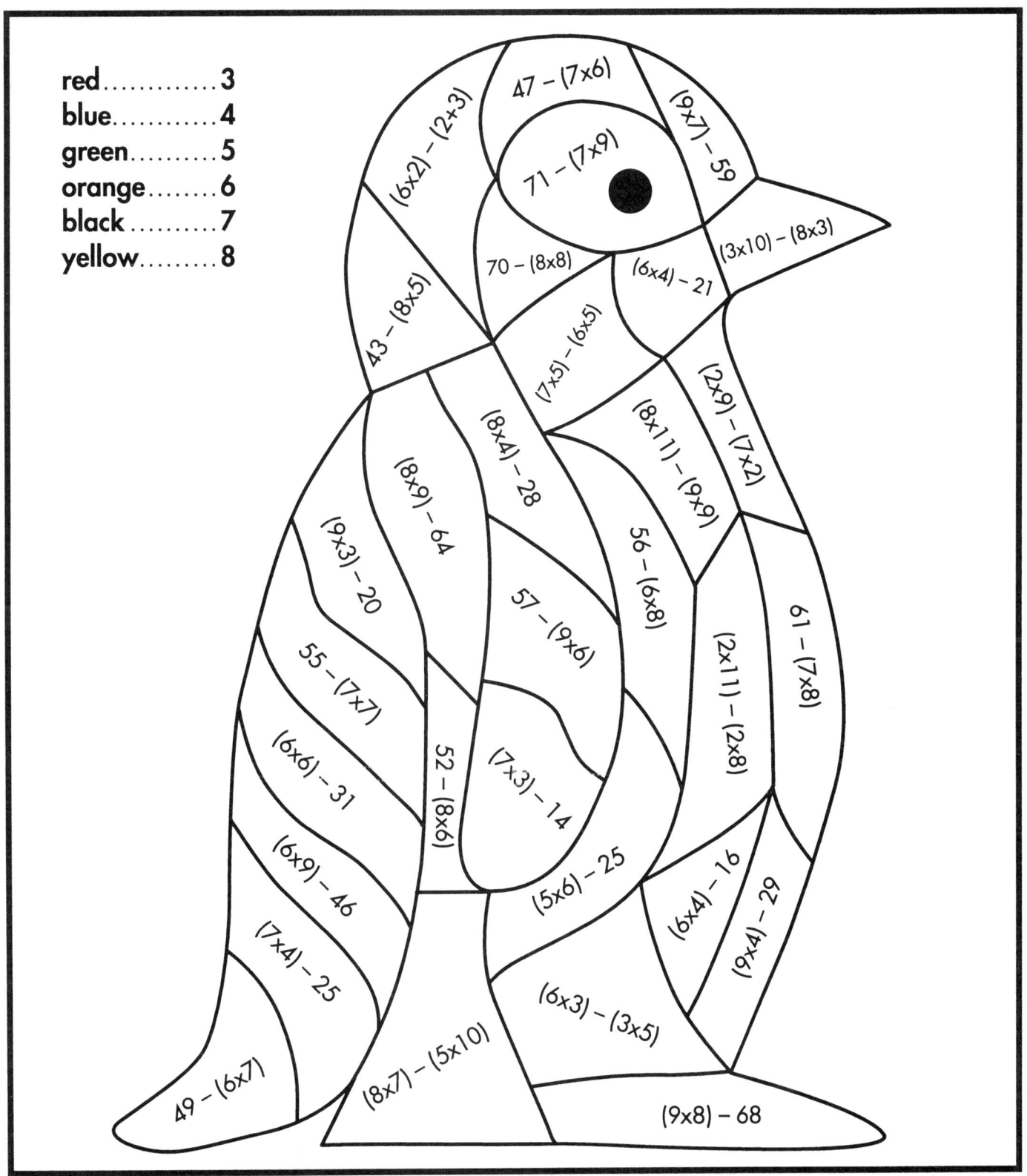

Correction Key

Pony

1. Allow your students to correct their own work.
2. Make a transparency of this puzzle and instruct your students to place the transparency over their completed puzzle for a quick and easy check.

Multiplication Puzzles ✦ Activity 39

Pony

Name _____

Date _____

1. Complete the problems within the parentheses first. Then complete the subtraction problem using your answer. (You can do the problems on another piece of paper.)

2. Using your final answer and the color key, color your puzzle correctly.

50 Total Problems

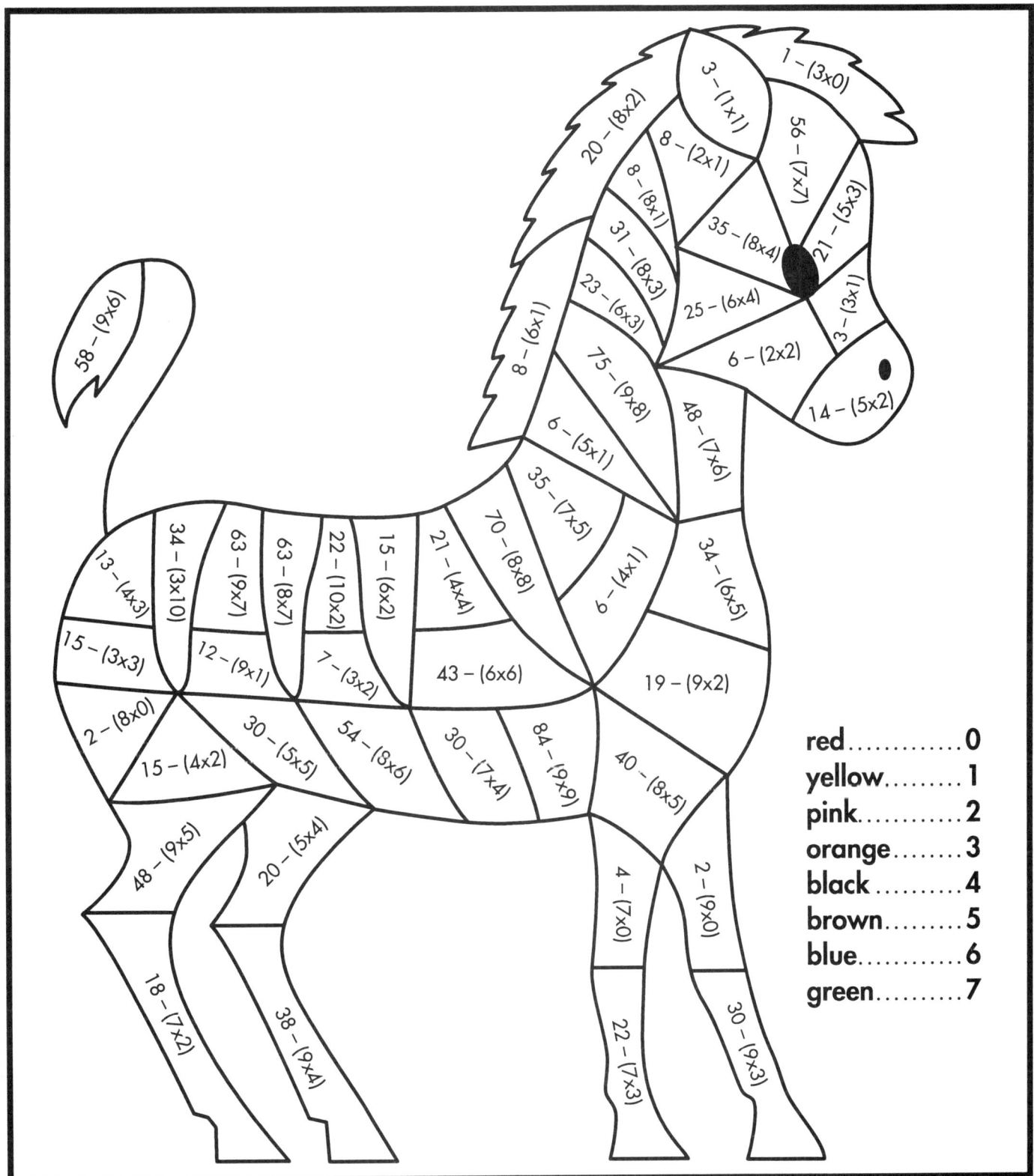

red............0
yellow........1
pink...........2
orange........3
black..........4
brown........5
blue...........6
green.........7

Multiplication Puzzles ♦ Activity 39

Correction Key

Poppy

1. Allow your students to correct their own work.
2. Make a transparency of this puzzle and instruct your students to place the transparency over their completed puzzle for a quick and easy check.

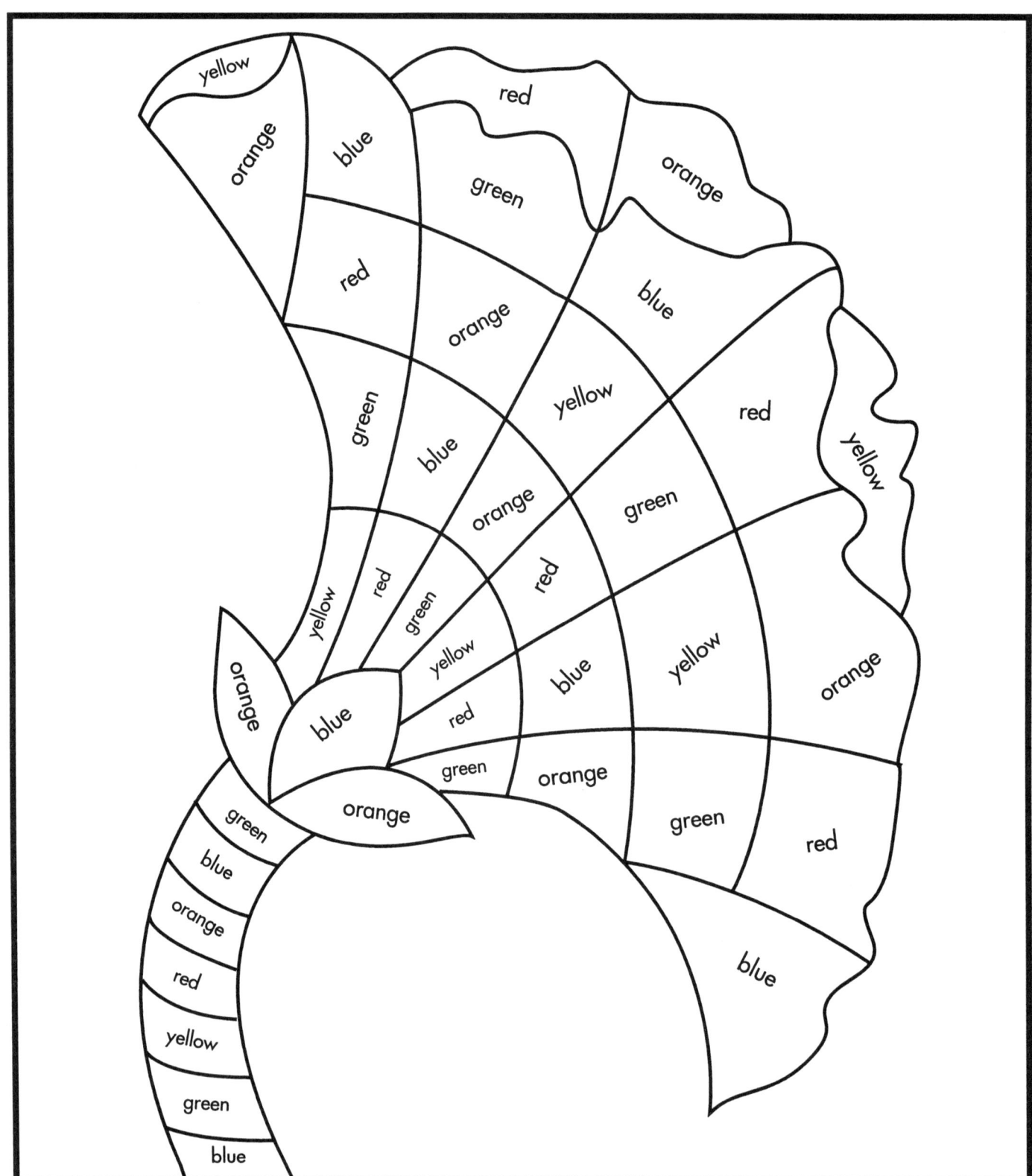

Multiplication Puzzles ✦ Activity 40

Poppy

Name _____

Date _____

1. Complete the problems within the parentheses first. Then complete the subtraction problem using your answer. (You can do the problems on another piece of paper.)

2. Using your final answer and the color key, color your puzzle correctly.

40 Total Problems

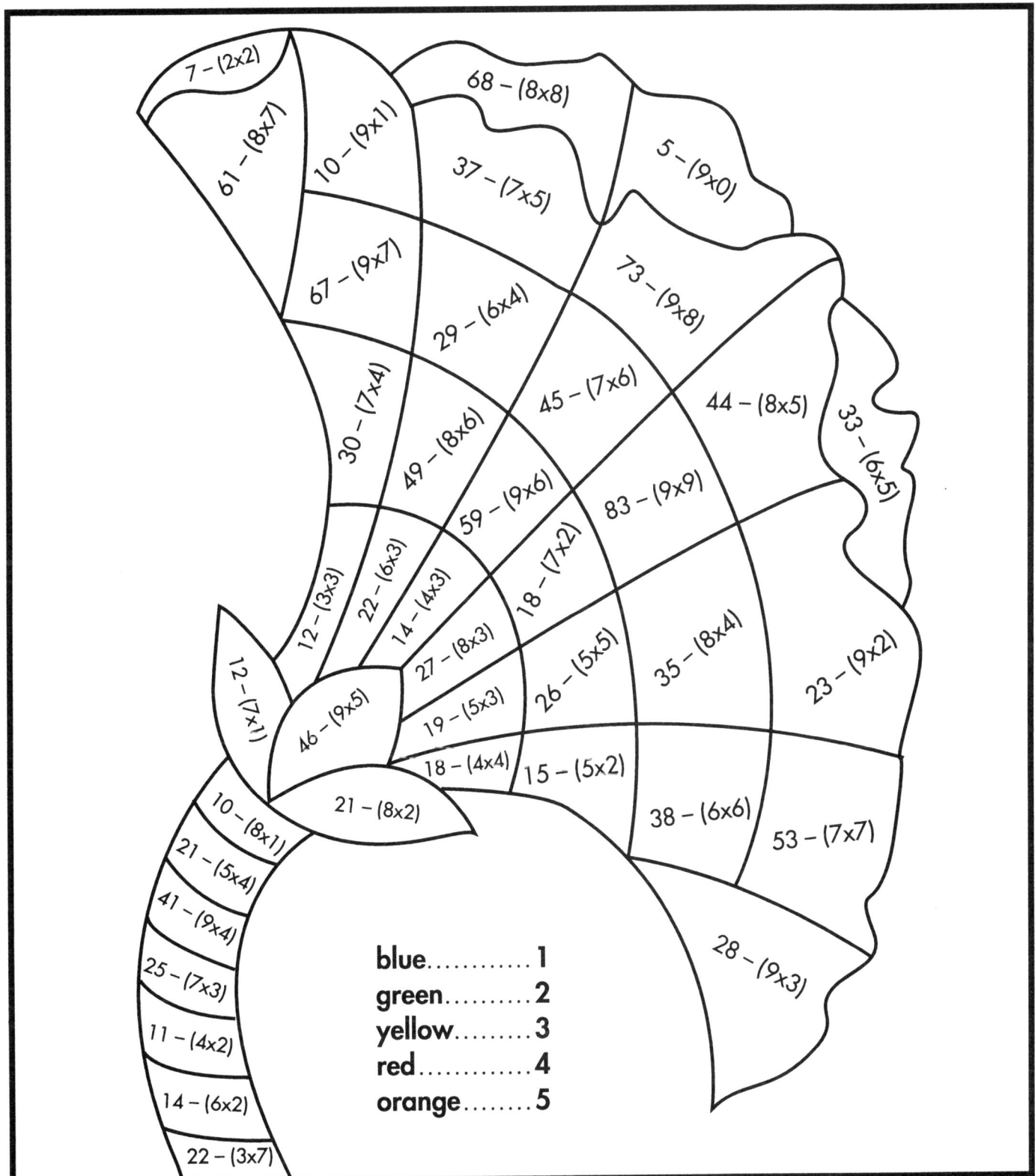

blue............1
green..........2
yellow........3
red............4
orange.......5

© Golden Educational Center

81

Multiplication Puzzles ♦ Activity 40

Correction Key

Rabbit

1. Allow your students to correct their own work.
2. Make a transparency of this puzzle and instruct your students to place the transparency over their completed puzzle for a quick and easy check.

Multiplication Puzzles ✦ Activity 41

Rabbit

Name _____

Date _____

1. Complete the problems within the parentheses first. Then complete the subtraction problem using your answer. (You can do the problems on another piece of paper.)
2. Using your final answer and the color key, color your puzzle correctly.

58 Total Problems

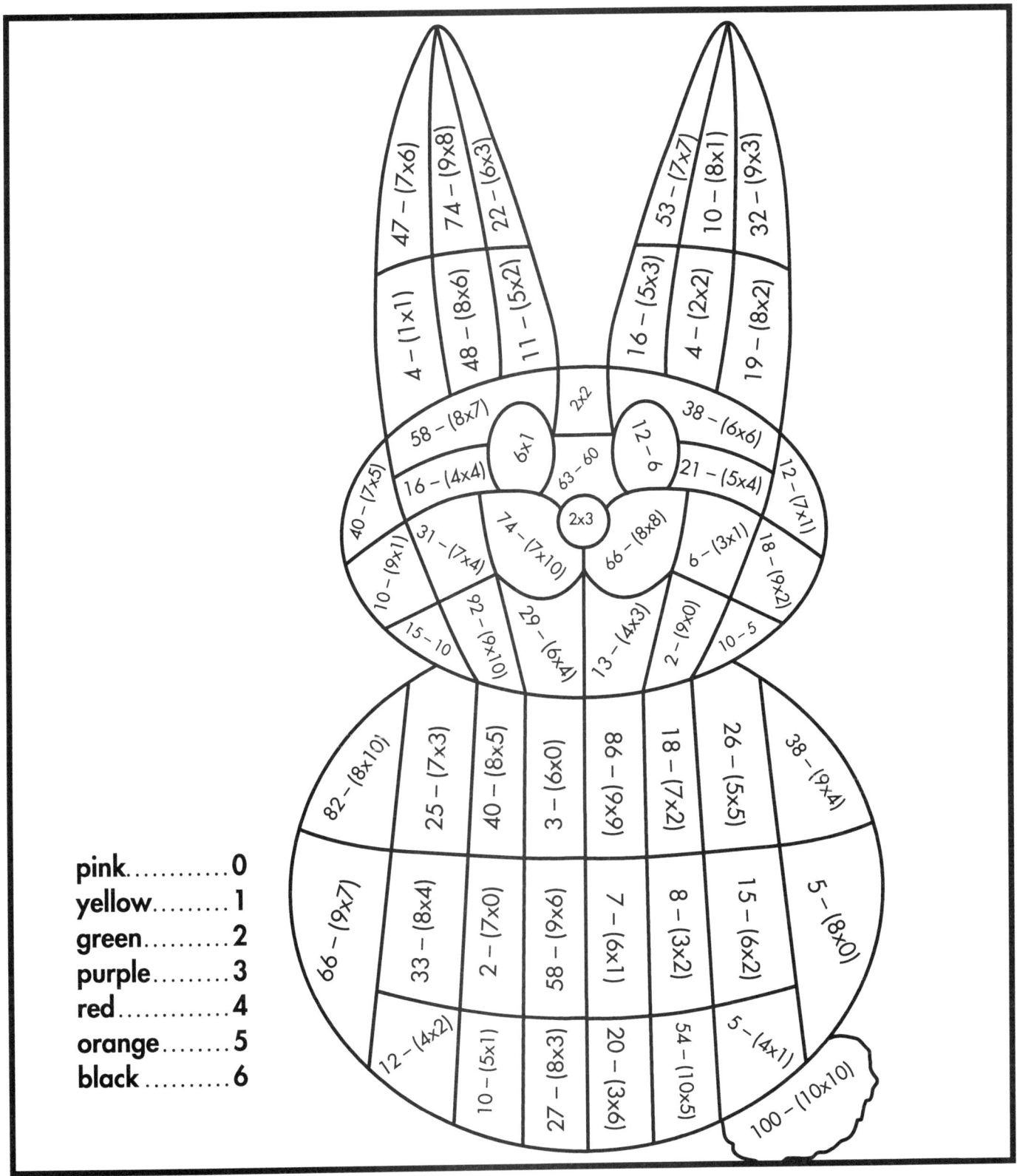

pink............0
yellow..........1
green..........2
purple..........3
red............4
orange..........5
black..........6

Multiplication Puzzles ✦ Activity 41

Correction Key

Raccoon

1. Allow your students to correct their own work.
2. Make a transparency of this puzzle and instruct your students to place the transparency over their completed puzzle for a quick and easy check.

Multiplication Puzzles ✦ Activity 42

Raccoon

Name _____

Date _____

1. Complete the problems within the parentheses first. Then complete the addition or subtraction problem using your answer. (You can do the problems on another piece of paper.)

2. Using your final answer and the color key, color your puzzle correctly.

54 Total Problems

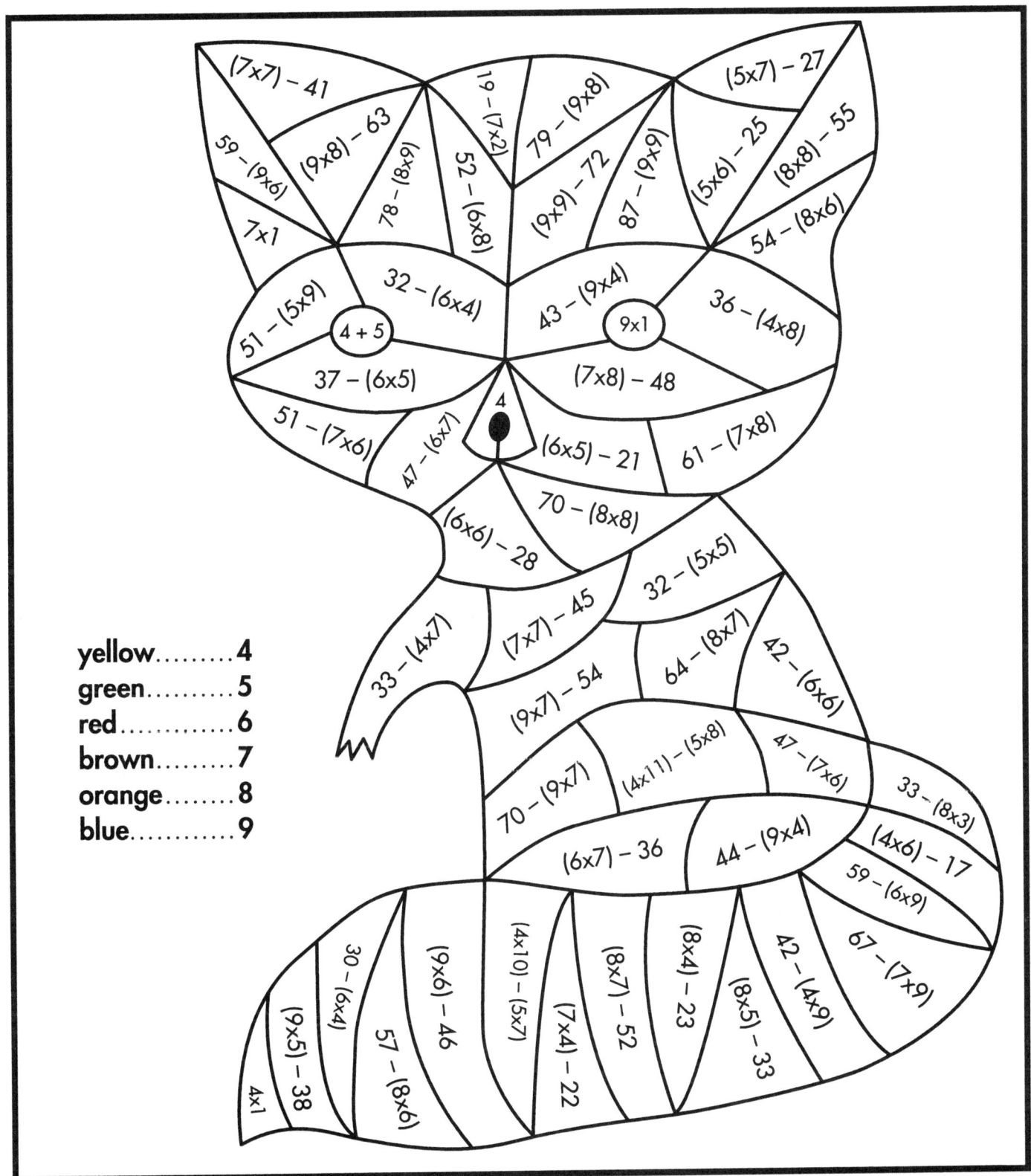

yellow........4
green.........5
red...........6
brown........7
orange.......8
blue..........9

Multiplication Puzzles ✦ Activity 42

Correction Key

1. Allow your students to correct their own work.
2. Make a transparency of this puzzle and instruct your students to place the transparency over their completed puzzle for a quick and easy check.

School Bus

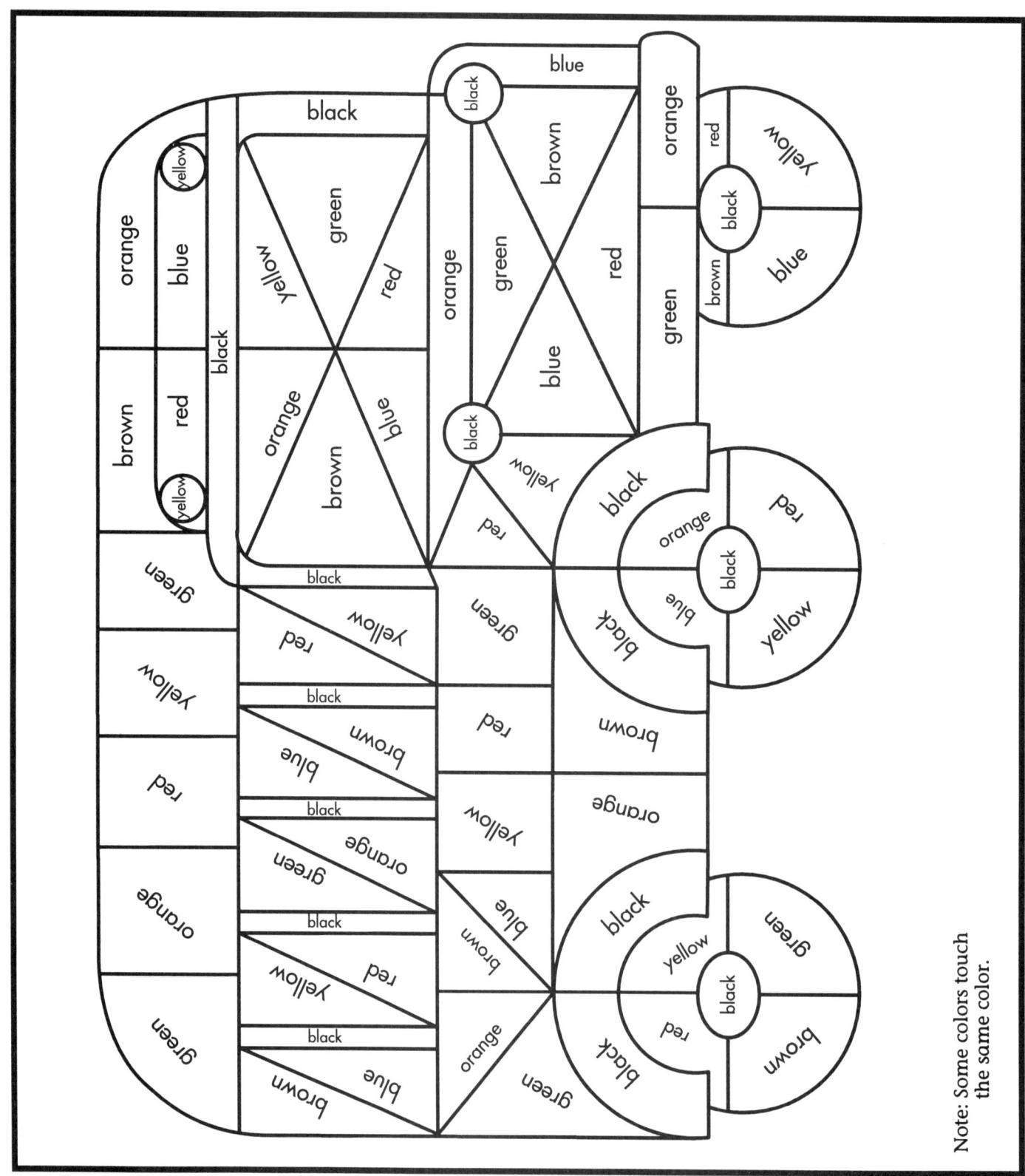

Multiplication Puzzles ♦ Activity 43

School Bus

1. Complete the problems within the parentheses first. Then complete the subtraction problem using your answer. (You can do the problems on another piece of paper.)
2. Using your final answer and the color key, color your puzzle correctly.

74 Total Problems

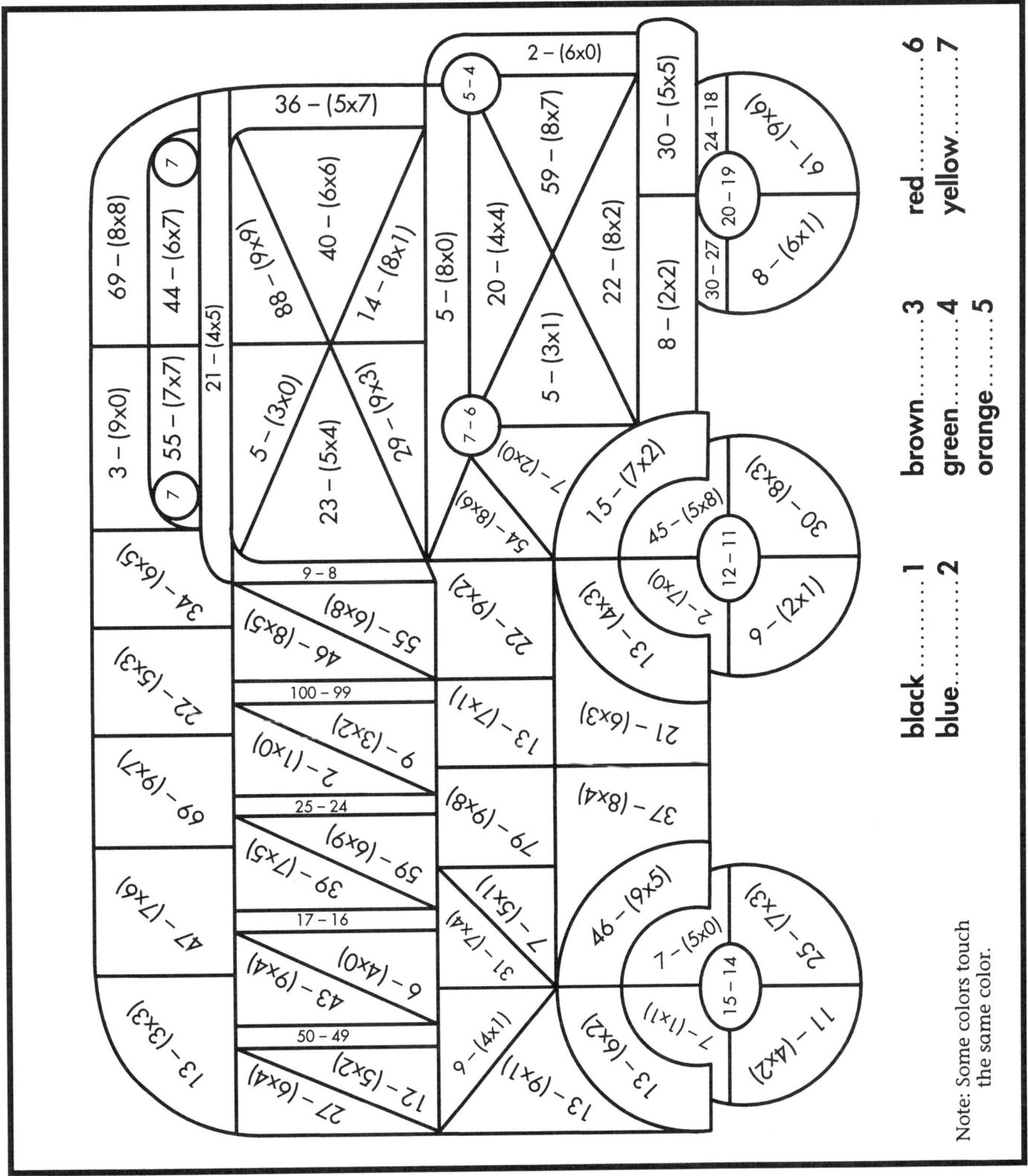

Multiplication Puzzles ✦ Activity 43

Correction Key

Scroll

1. Allow your students to correct their own work.
2. Make a transparency of this puzzle and instruct your students to place the transparency over their completed puzzle for a quick and easy check.

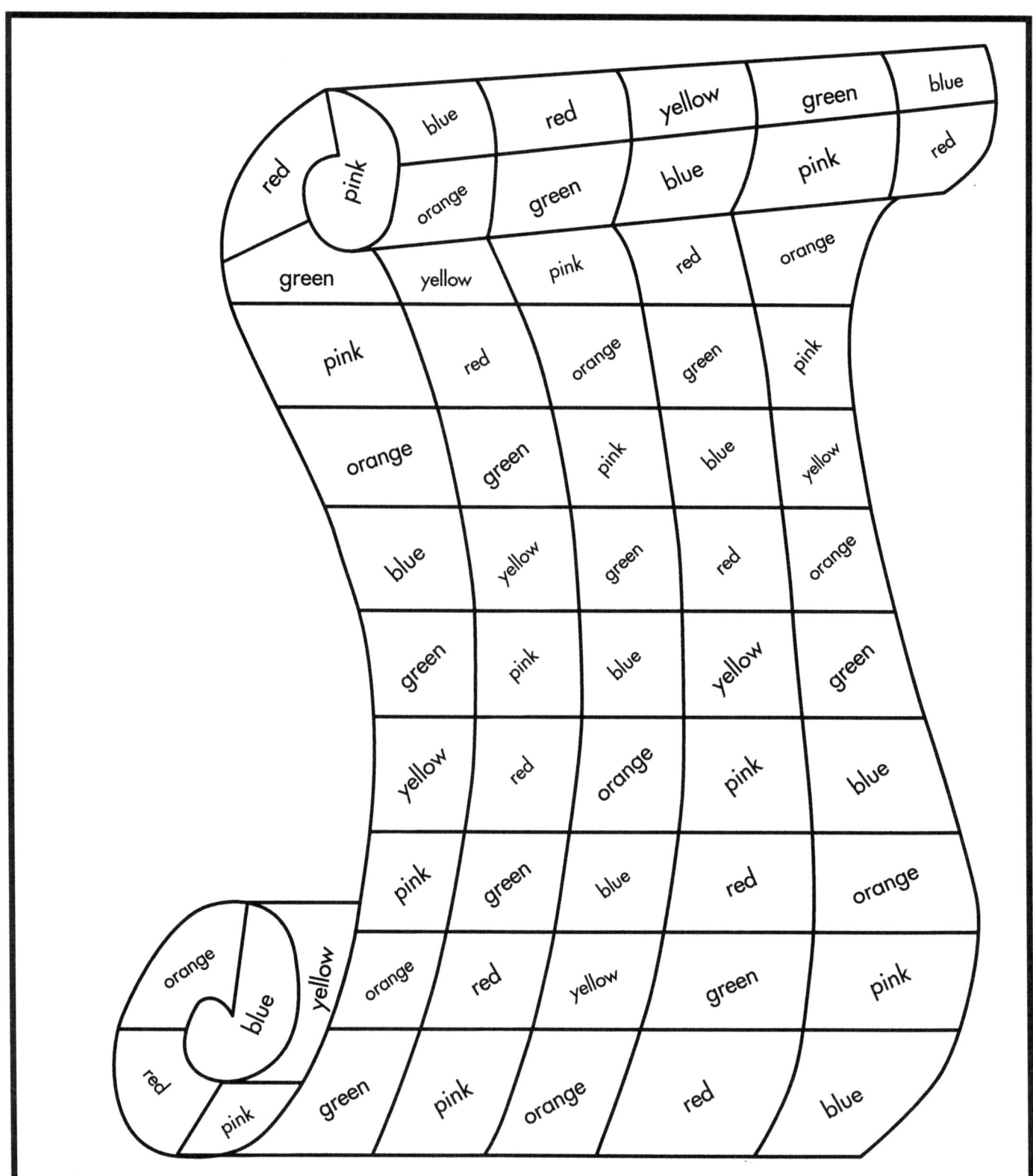

Multiplication Puzzles ✦ Activity 44

Scroll

Name _____

Date _____

1. Complete the problems within the parentheses first. Then complete the subtraction problem using your answer. (You can do the problems on another piece of paper.)

2. Using your final answer and the color key, color your puzzle correctly.

62 Total Problems

Color Key:
- blue........... 1
- green.......... 2
- orange........ 3
- pink........... 4
- red............ 5
- yellow......... 6

Puzzle cells:

- 25 – (4x5)
- 16 – (3x4)
- 13 – (6x2)
- 32 – (9x3)
- 18 – (4x3)
- 30 – (7x4)
- 41 – (8x5)
- 35 – (8x4)
- 2 – (2x0)
- 82 – (9x9)
- 22 – (6x3)
- 45 – (10x4)
- 18 – (4x4)
- 42 – (9x4)
- 39 – (7x5)
- 25 – (5x4)
- 51 – (8x6)
- 28 – (8x3)
- 26 – (7x3)
- 21 – (9x2)
- 2 – (5x0)
- 28 – (6x4)
- 17 – (7x2)
- 2 – (0x1)
- 29 – (5x5)
- 43 – (7x6)
- 78 – (9x8)
- 1 – (6x0)
- 36 – (10x3)
- 11 – (3x3)
- 13 – (4x2)
- 59 – (8x7)
- 2 – (4x0)
- 13 – (9x1)
- 64 – (9x7)
- 8 – (2x1)
- 7 – (5x1)
- 22 – (8x2)
- 55 – (10x5)
- 52 – (7x7)
- 10 – (3x2)
- 65 – (8x8)
- 8 – (4x1)
- 56 – (9x6)
- 21 – (10x2)
- 5 – (7x0)
- 9 – (6x1)
- 39 – (6x6)
- 8 – (7x1)
- 42 – (6x6)
- 11 – (8x1)
- 9 – (2x2)
- 51 – (9x5)
- 32 – (6x5)
- 4 – (9x0)
- 15 – (5x2)
- 60 – (7x8)
- 2 – (3x0)
- 5 – (1x1)
- 18 – (5x3)
- 8 – (3x1)
- 11 – (10x1)

Multiplication Puzzles ✦ Activity 44

Correction Key   **Seal**

1. Allow your students to correct their own work.
2. Make a transparency of this puzzle and instruct your students to place the transparency over their completed puzzle for a quick and easy check.

Multiplication Puzzles ✦ Activity 45

Seal

Name _____

Date _____

1. Complete the problems within the parentheses first. Then complete the subtraction problem using your answer. (You can do the problems on another piece of paper.)
2. Using your final answer and the color key, color your puzzle correctly.

37 Total Problems

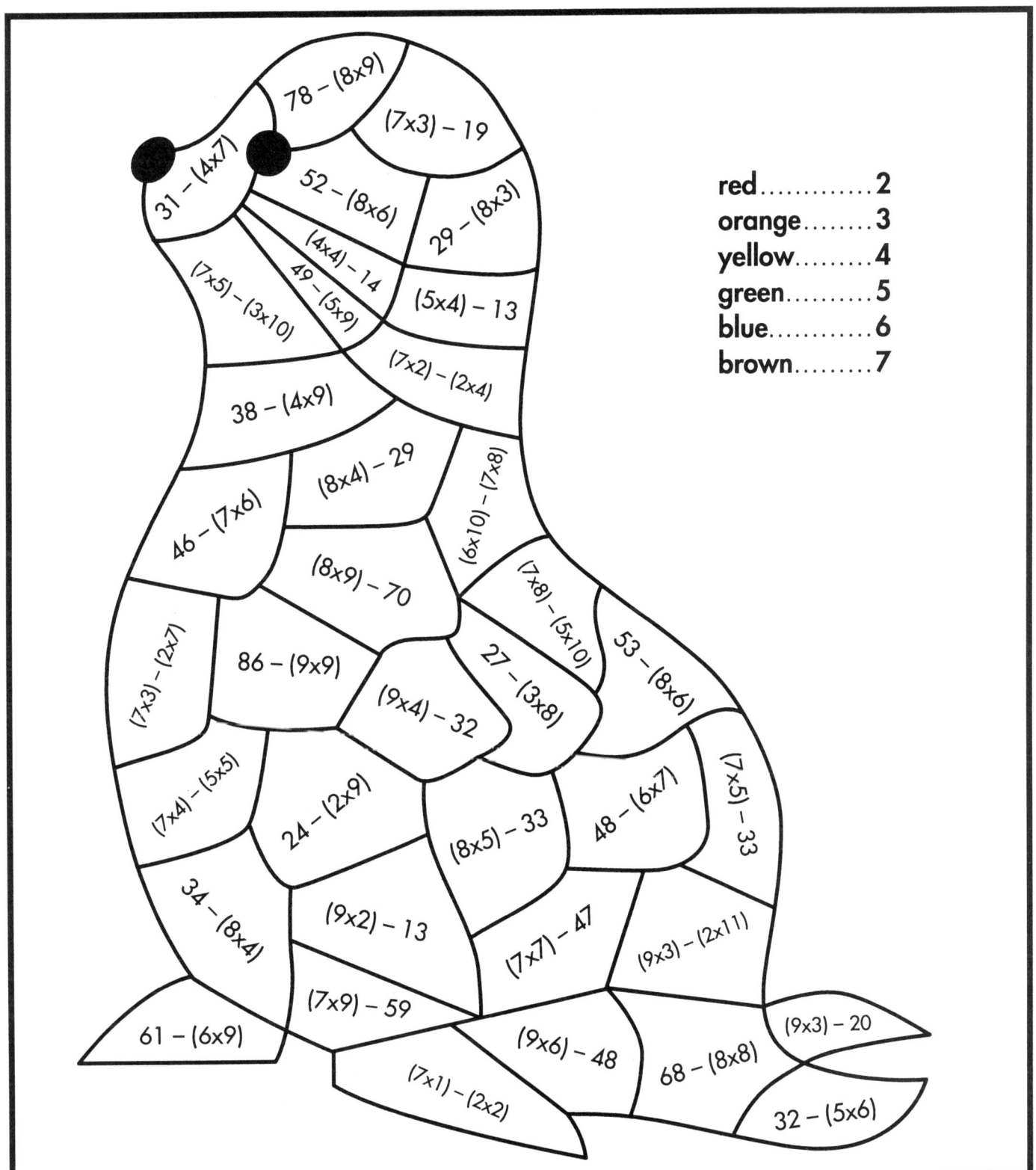

red............2
orange........3
yellow........4
green..........5
blue............6
brown........7

Multiplication Puzzles ♦ Activity 45

Correction Key

Shamrock

1. Allow your students to correct their own work.
2. Make a transparency of this puzzle and instruct your students to place the transparency over their completed puzzle for a quick and easy check.

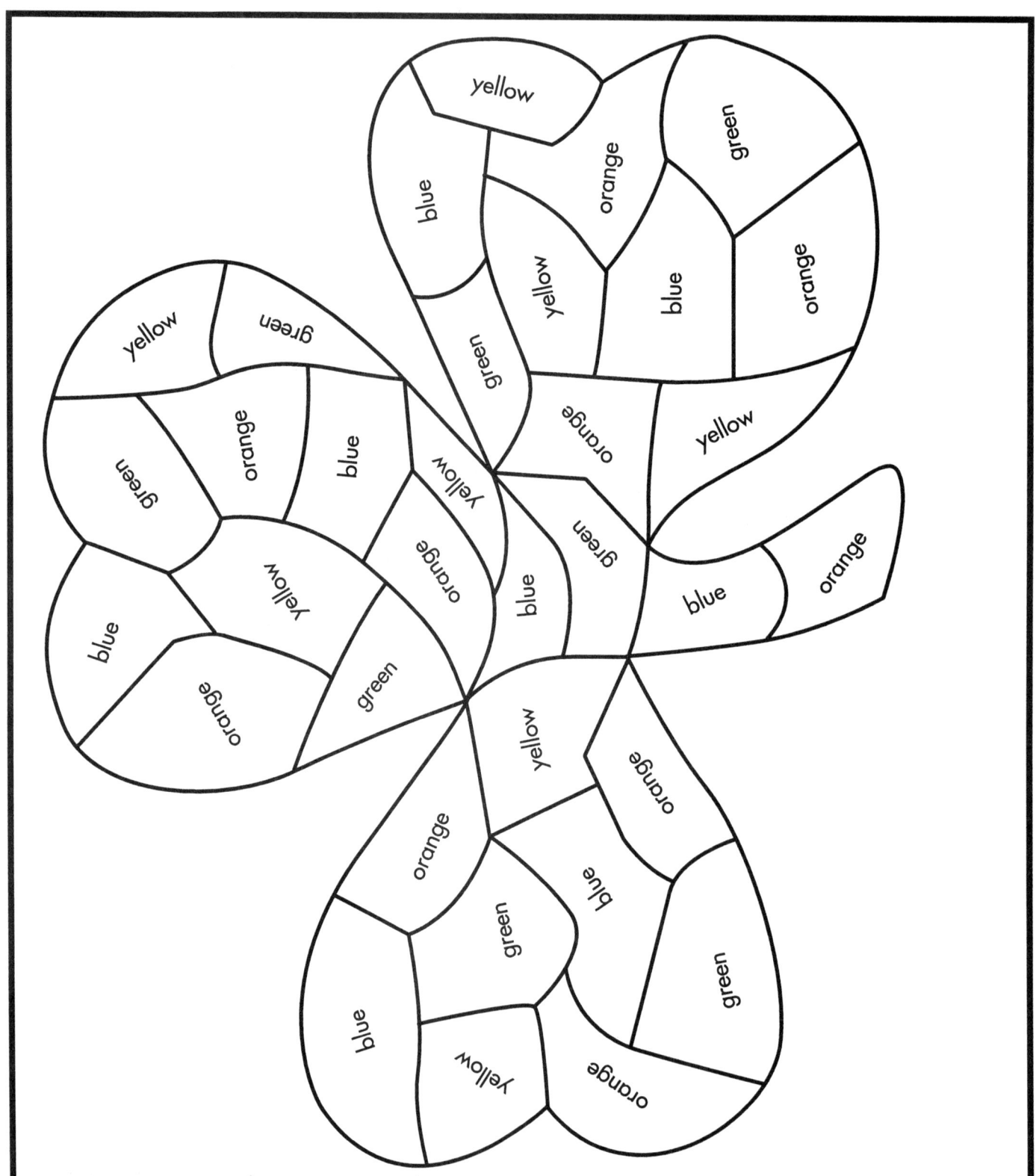

Multiplication Puzzles ✦ Activity 46

Shamrock

Name _____

Date _____

1. Complete the problems within the parentheses first. Then complete the subtraction problem using your answer. (You can do the problems on another piece of paper.)
2. Using your final answer and the color key, color your puzzle correctly.

34 Total Problems

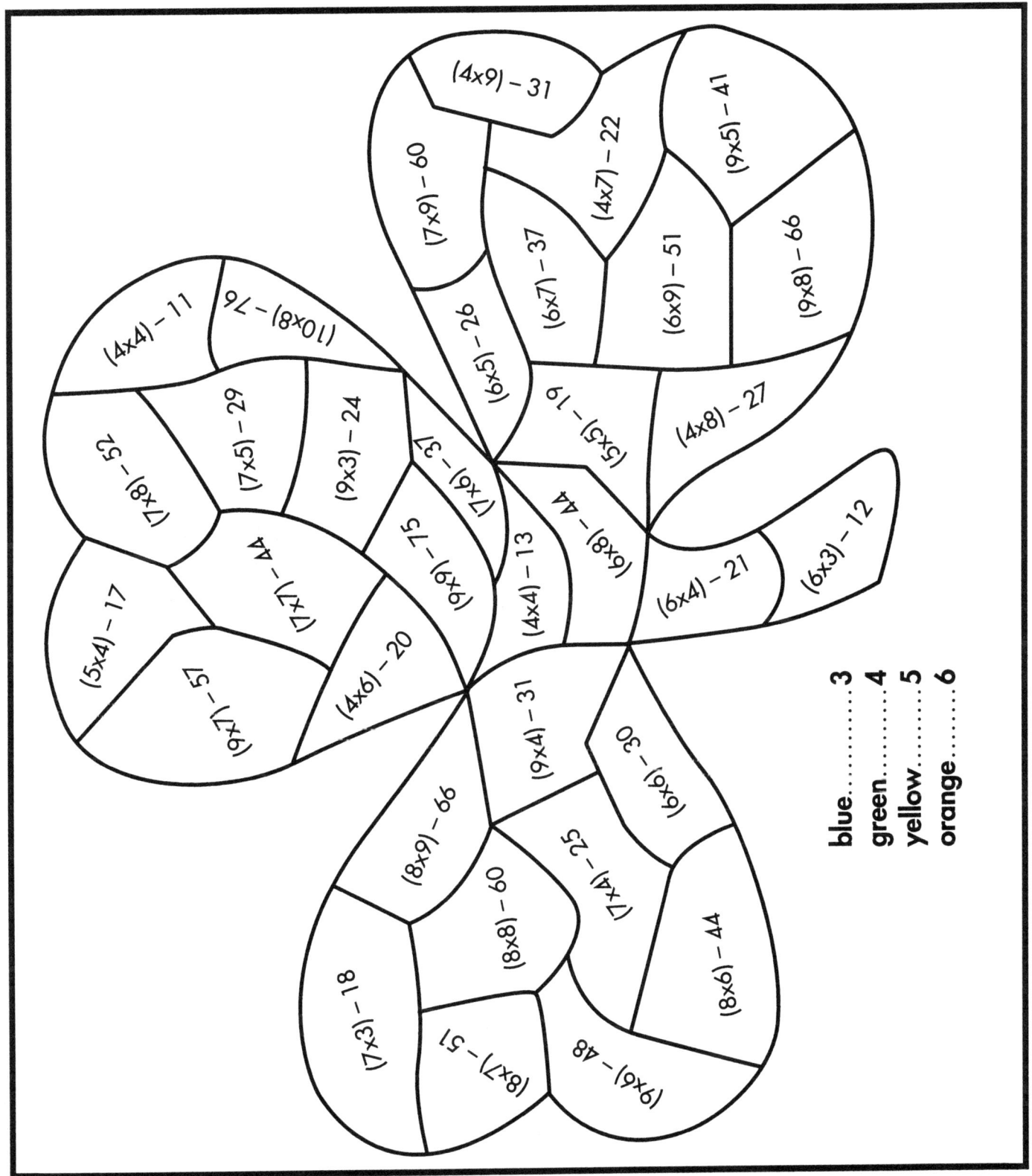

Multiplication Puzzles ✦ Activity 46

Correction Key

Snowman

1. Allow your students to correct their own work.
2. Make a transparency of this puzzle and instruct your students to place the transparency over their completed puzzle for a quick and easy check.

Multiplication Puzzles ✦ Activity 47

Snowman

Name _____

Date _____

1. Complete the problems within the parentheses first. Then complete the subtraction problem using your answer. (You can do the problems on another piece of paper.)

2. Using your final answer and the color key, color your puzzle correctly.

42 Total Problems

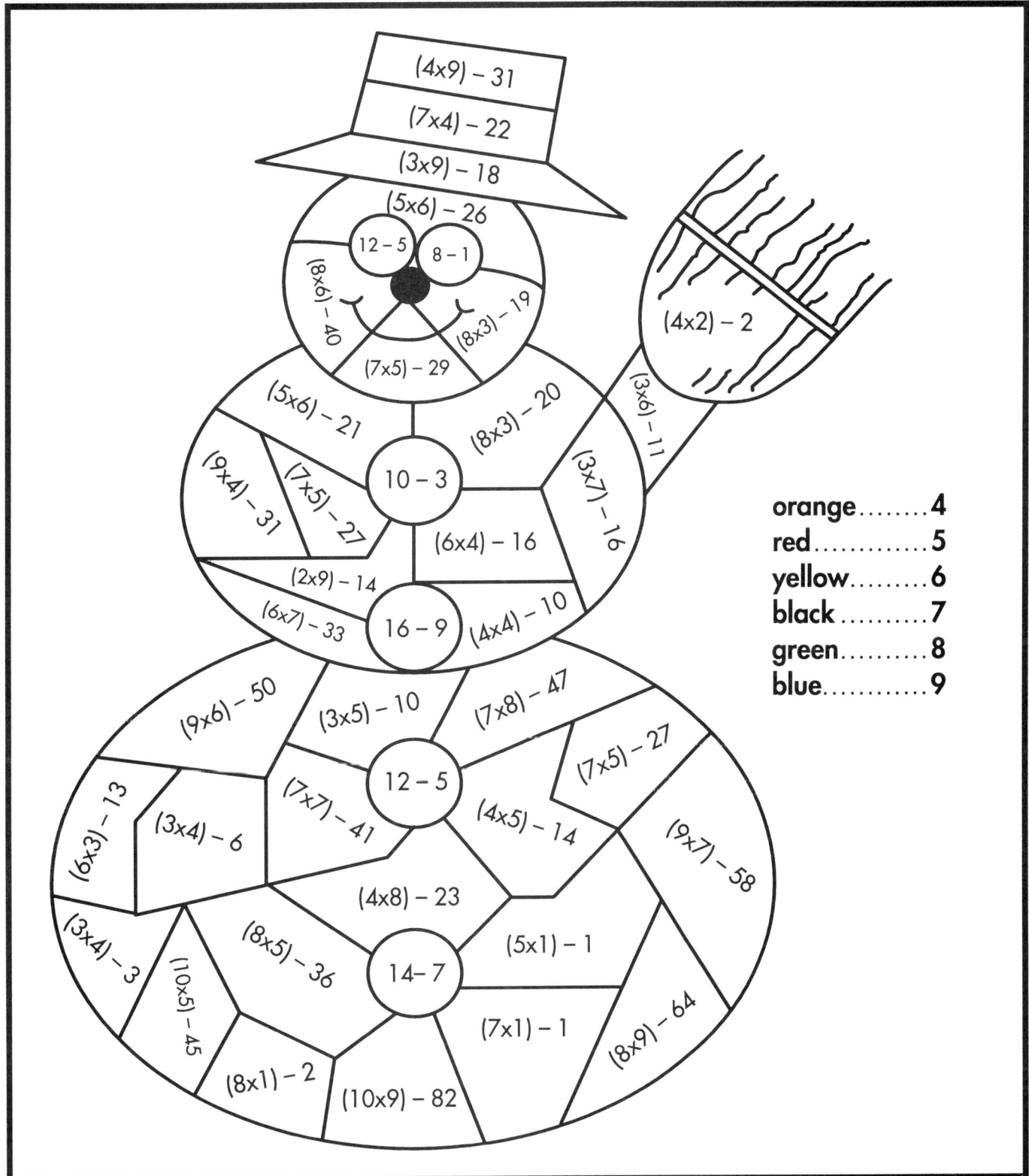

orange........4
red.............5
yellow........6
black..........7
green..........8
blue............9

Multiplication Puzzles ♦ Activity 47

Correction Key

Squirrel

1. Allow your students to correct their own work.
2. Make a transparency of this puzzle and instruct your students to place the transparency over their completed puzzle for a quick and easy check.

Multiplication Puzzles ✦ Activity 48

© Golden Educational Center

Squirrel

Name _____

Date _____

1. Complete the problems within the parentheses first. Then complete the subtraction problem using your answer. (You can do the problems on another piece of paper.)
2. Using your final answer and the color key, color your puzzle correctly.

55 Total Problems

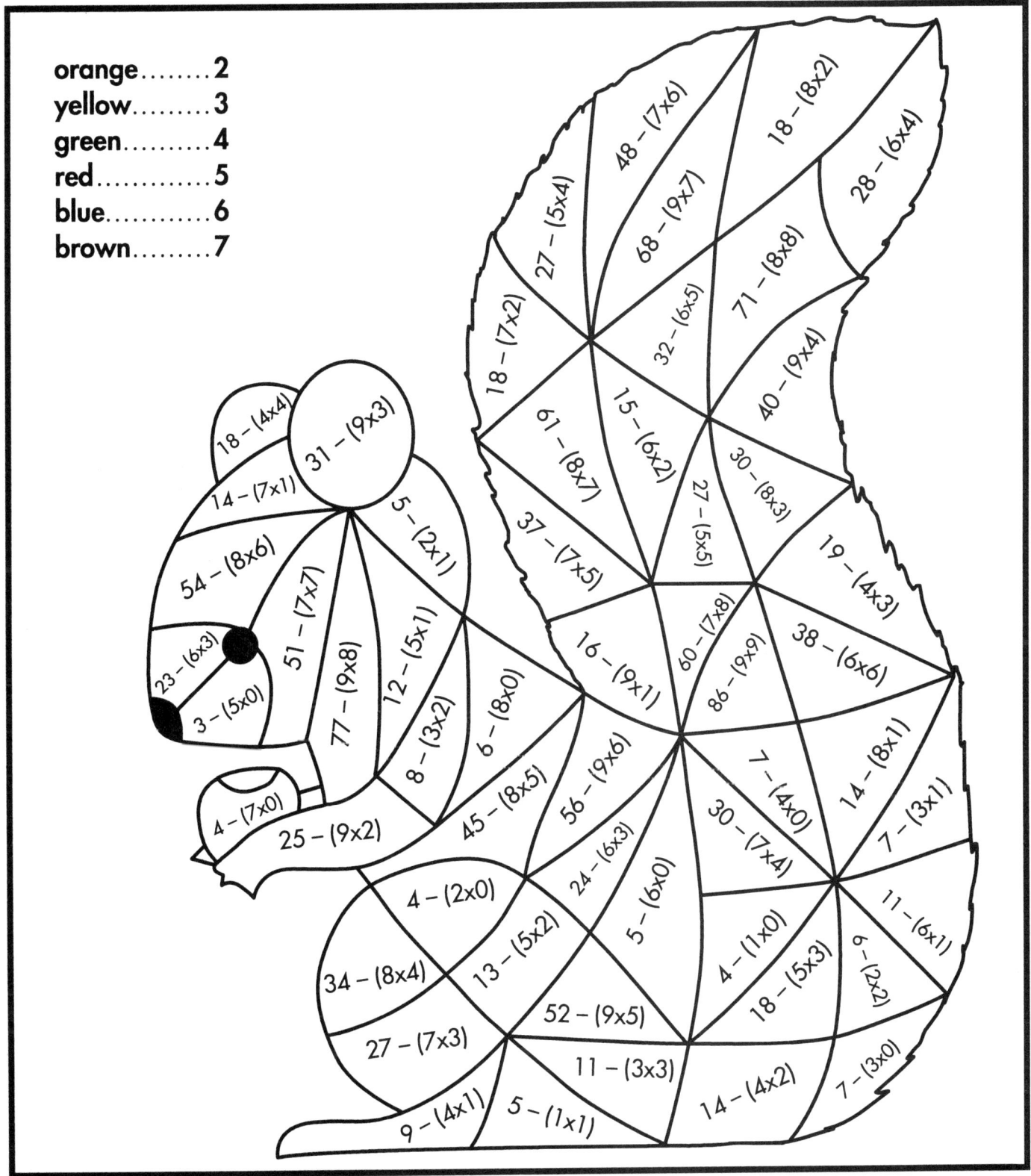

orange........2
yellow........3
green.........4
red............5
blue...........6
brown........7

© Golden Educational Center

Multiplication Puzzles ✦ Activity 48

Correction Key

Star

1. Allow your students to correct their own work.
2. Make a transparency of this puzzle and instruct your students to place the transparency over their completed puzzle for a quick and easy check.

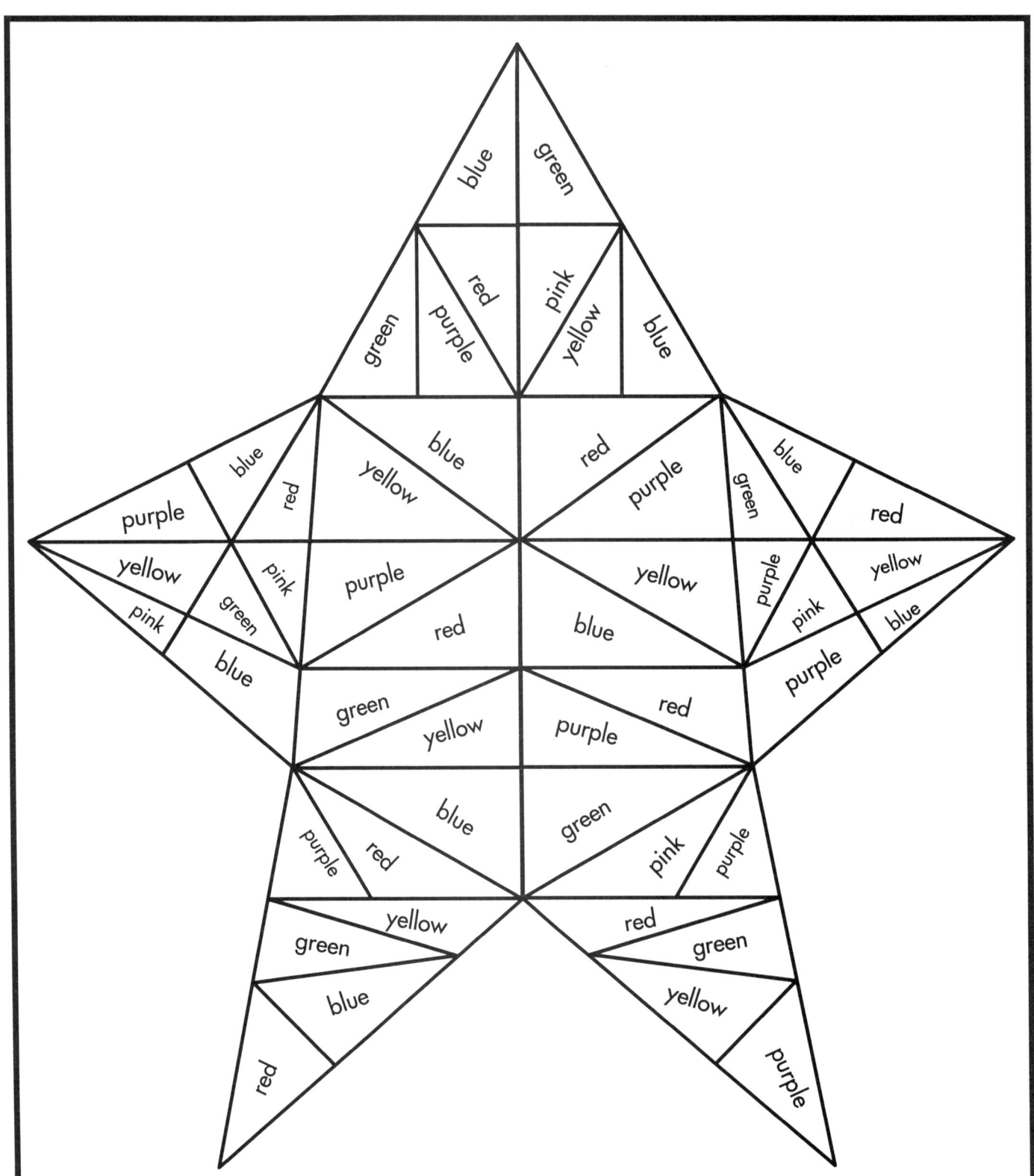

Multiplication Puzzles ✦ Activity 49

Star

Name _____

Date _____

1. Complete the problems within the parentheses first. Then complete the subtraction problem using your answer. (You can do the problems on another piece of paper.)
2. Using your final answer and the color key, color your puzzle correctly.

50 Total Problems

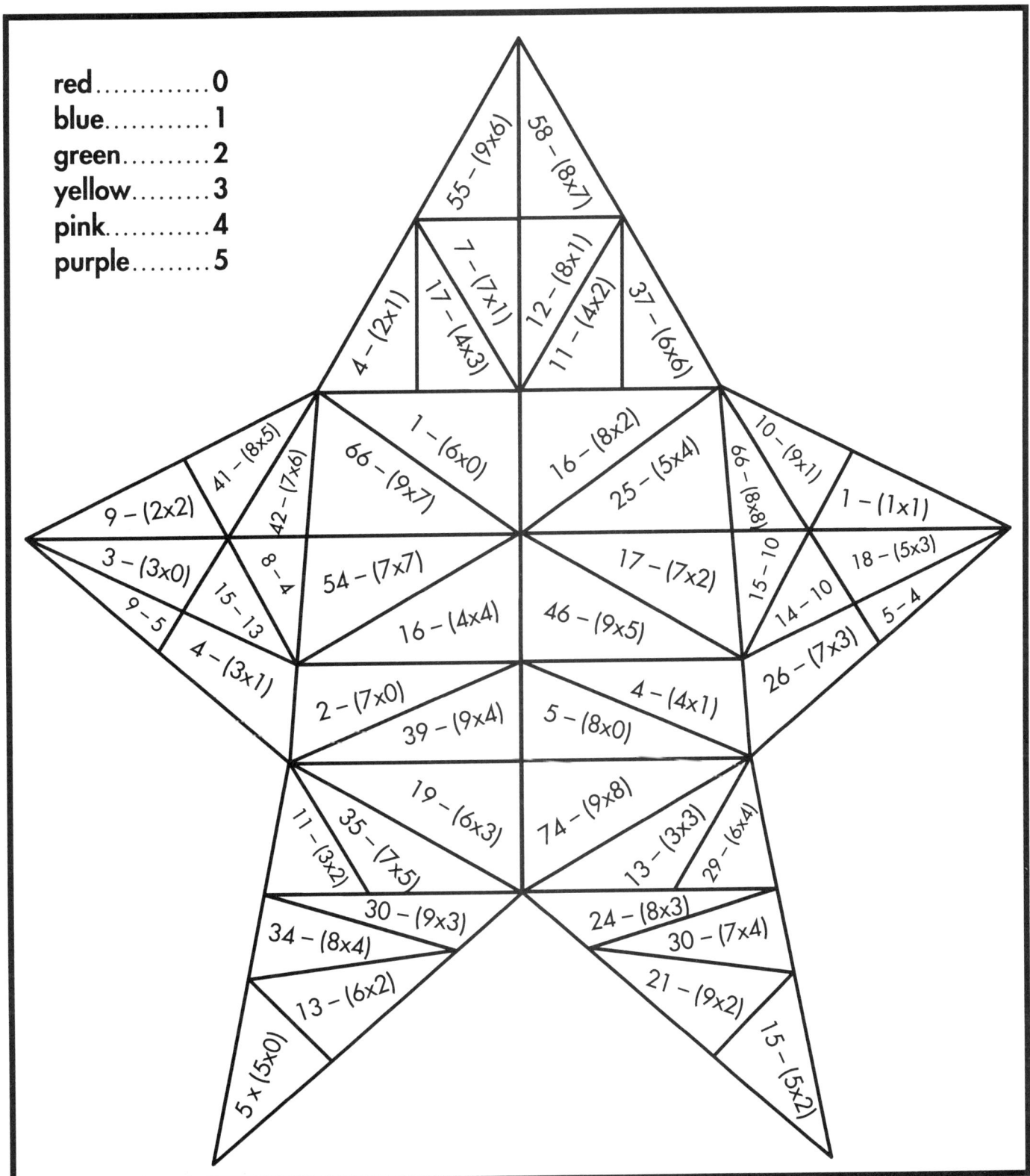

Multiplication Puzzles ♦ Activity 49

Correction Key

Statue of Liberty

1. Allow your students to correct their own work.
2. Make a transparency of this puzzle and instruct your students to place the transparency over their completed puzzle for a quick and easy check.

Multiplication Puzzles ✦ Activity 50

Statue of Liberty

Name _____

Date _____

1. Complete the problems within the parentheses first. Then complete the subtraction problem using your answer. (You can do the problems on another piece of paper.)
2. Using your final answer and the color key, color your puzzle correctly.

59 Total Problems

yellow.........1
green..........2
orange........3
brown........4
red............5
blue...........6

Multiplication Puzzles ♦ Activity 50

Correction Key

Stocking

1. Allow your students to correct their own work.
2. Make a transparency of this puzzle and instruct your students to place the transparency over their completed puzzle for a quick and easy check.

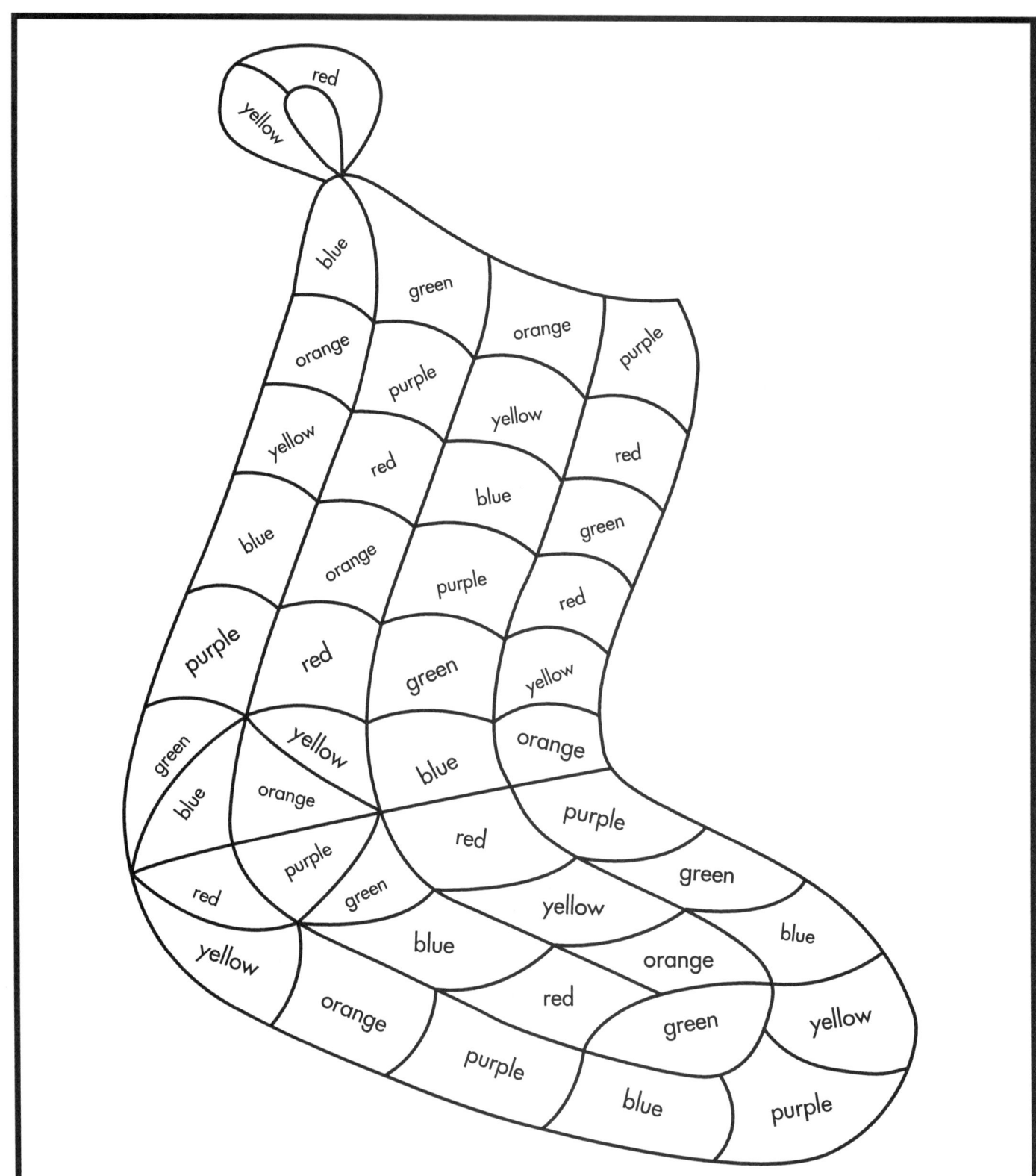

Multiplication Puzzles ♦ Activity 51

Stocking

Name _____

Date _____

1. Complete the problems within the parentheses first. Then complete the subtraction problem using your answer. (You can do the problems on another piece of paper.)
2. Using your final answer and the color key, color your puzzle correctly.

46 Total Problems

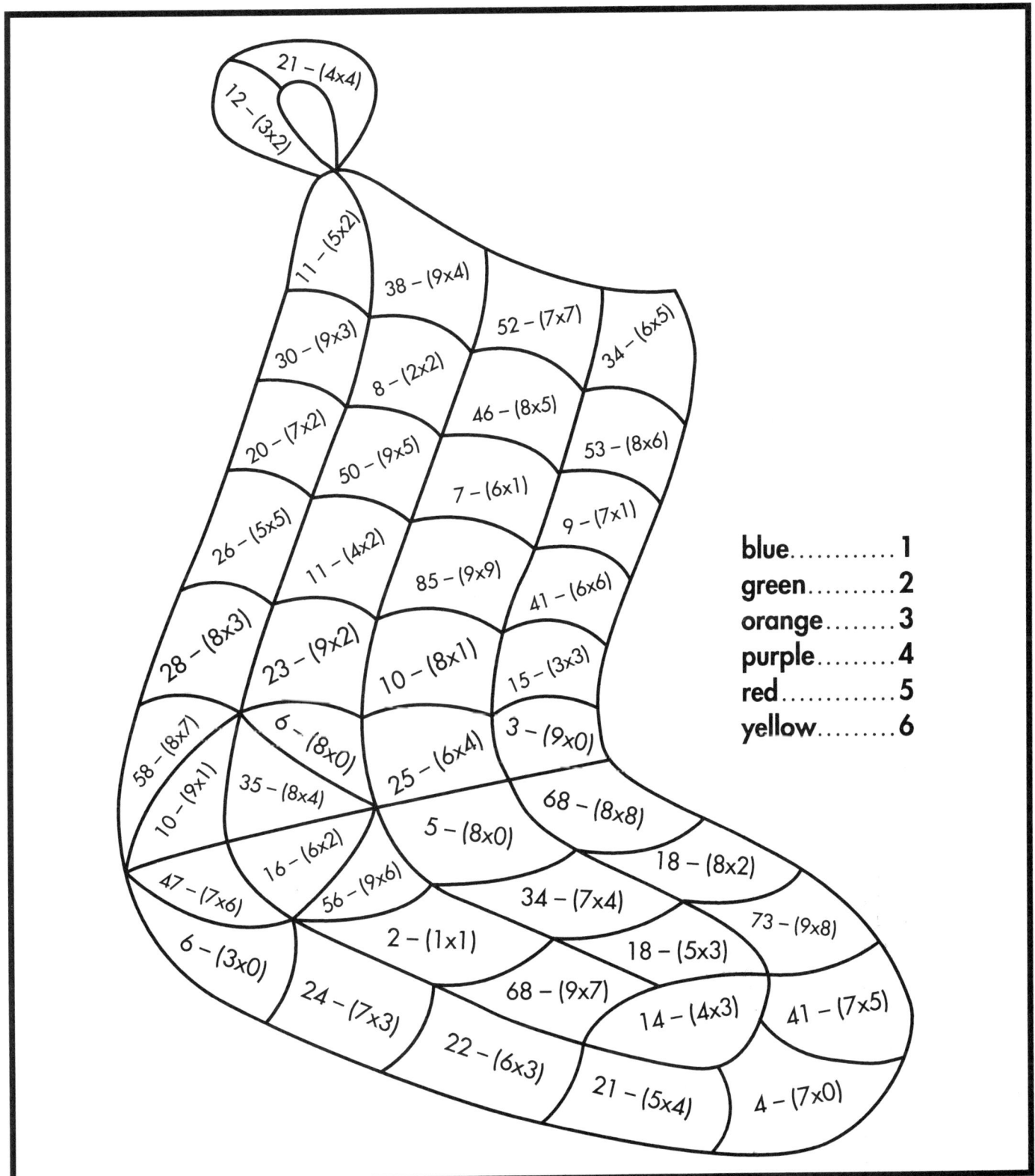

blue............1
green..........2
orange........3
purple.........4
red.............5
yellow.........6

Multiplication Puzzles ✦ Activity 51

Correction Key

Strawberry

1. Allow your students to correct their own work.
2. Make a transparency of this puzzle and instruct your students to place the transparency over their completed puzzle for a quick and easy check.

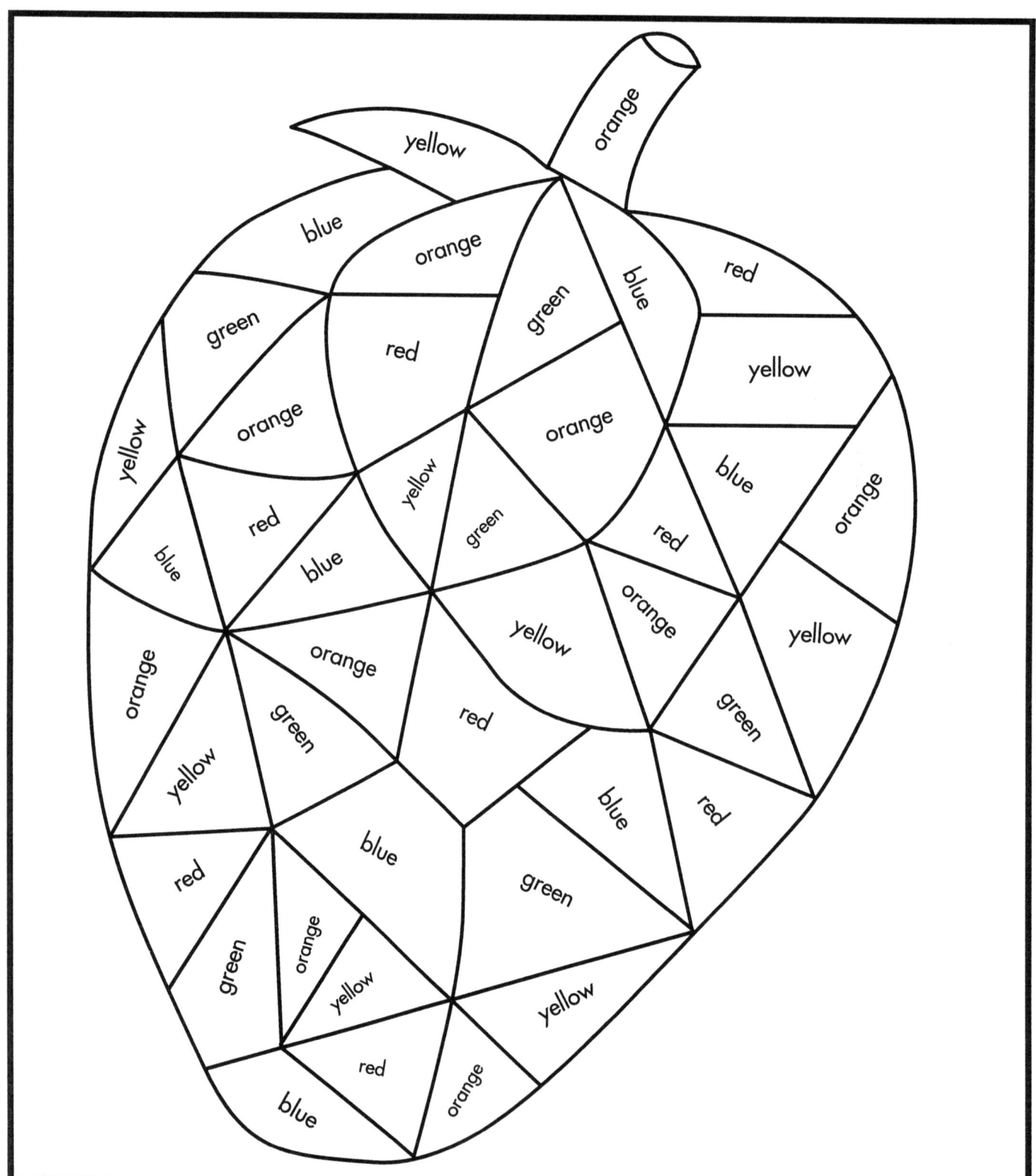

Multiplication Puzzles ✦ Activity 52

Strawberry

Name _____

Date _____

1. Complete the problems within the parentheses first. Then complete the subtraction problem using your answer. (You can do the problems on another piece of paper.)

2. Using your final answer and the color key, color your puzzle correctly.

42 Total Problems

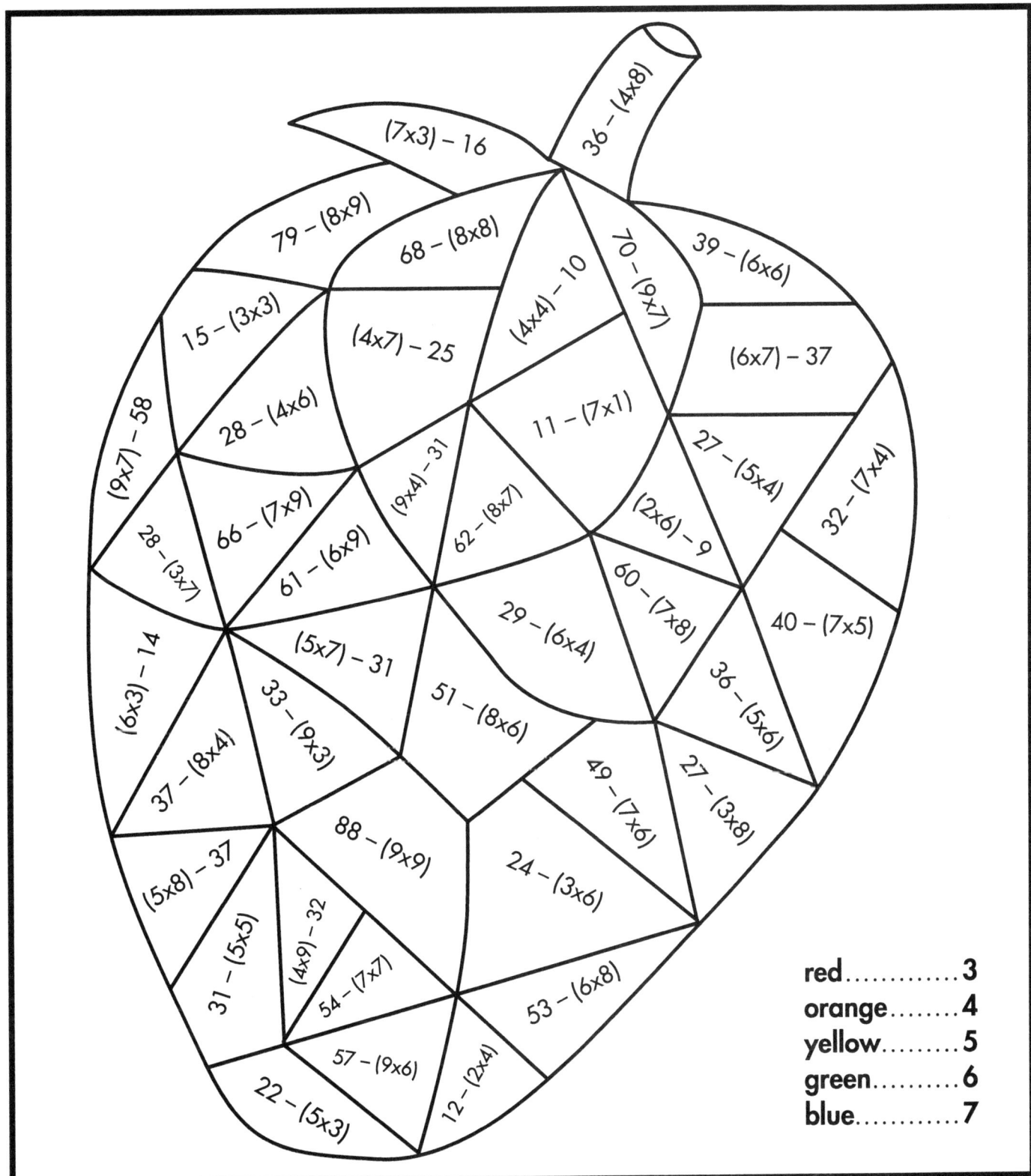

red..........3
orange......4
yellow......5
green.......6
blue........7

Multiplication Puzzles ✦ Activity 52

Correction Key

Sunflower

1. Allow your students to correct their own work.
2. Make a transparency of this puzzle and instruct your students to place the transparency over their completed puzzle for a quick and easy check.

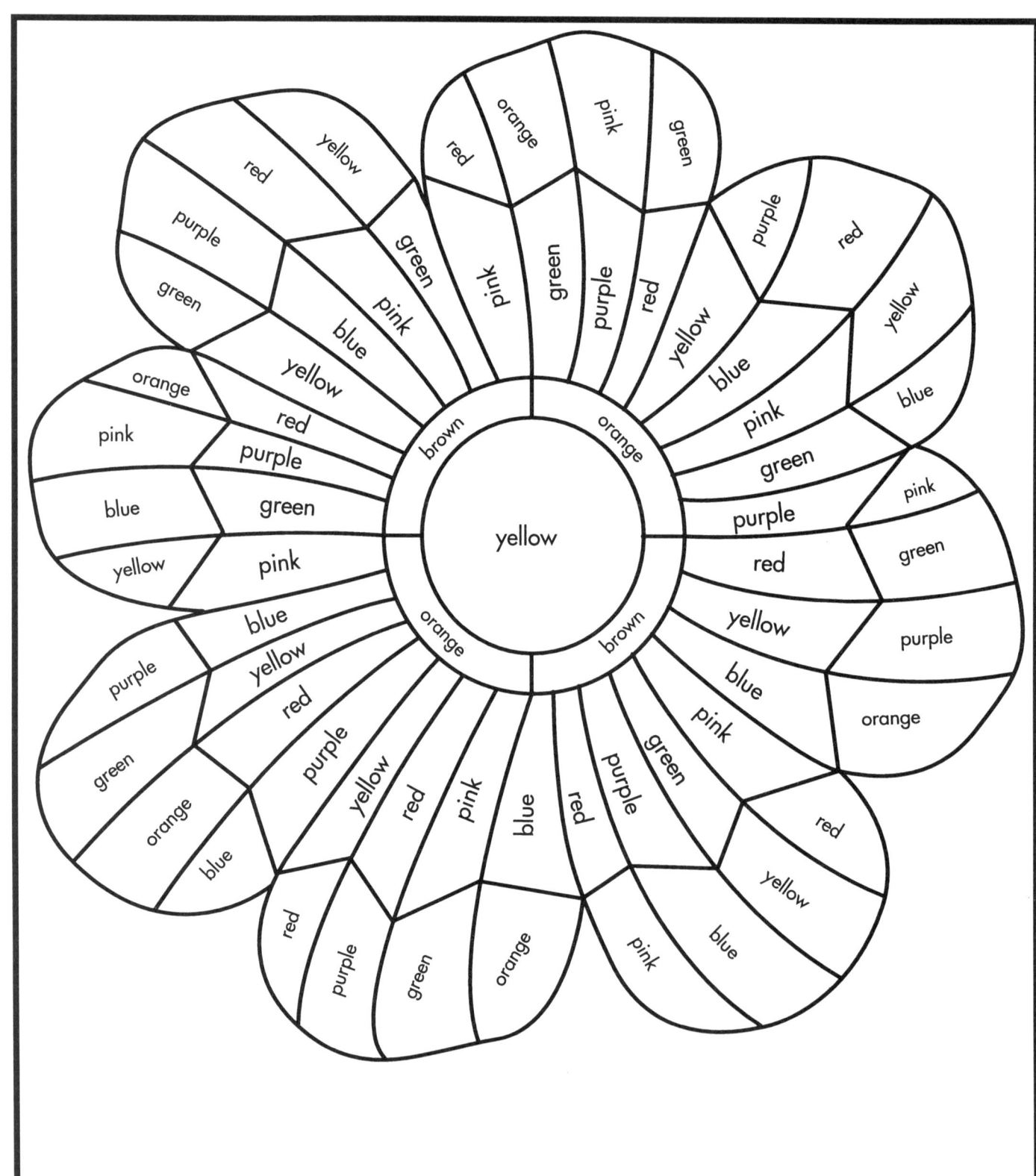

Multiplication Puzzles ✦ Activity 53

Sunflower

Name _____

Date _____

1. Complete the problems within the parentheses first. Then complete the subtraction problem using your answer. (You can do the problems on another piece of paper.)
2. Using your final answer and the color key, color your puzzle correctly.

69 Total Problems

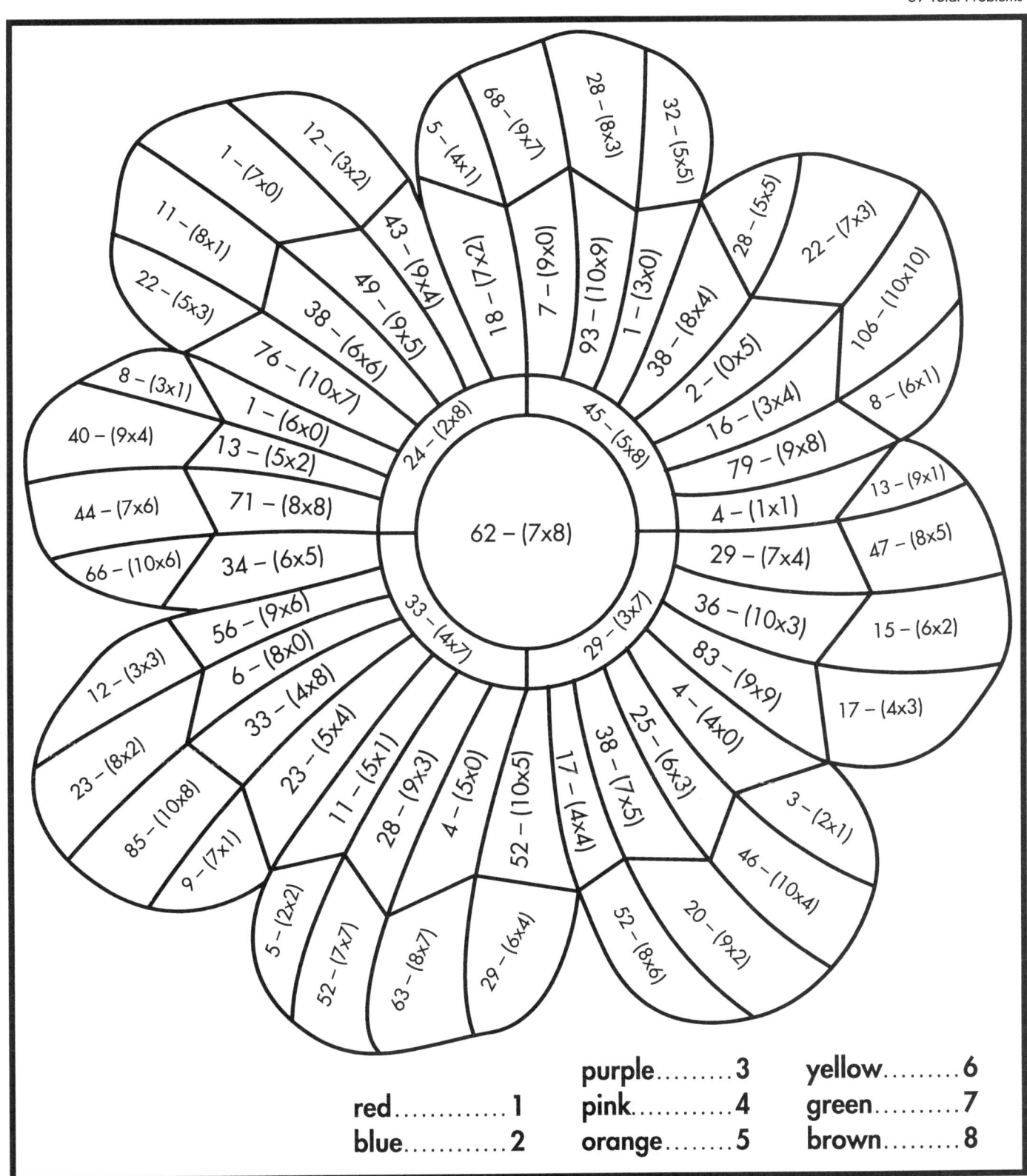

red............1
blue............2
purple............3
pink............4
orange............5
yellow............6
green............7
brown............8

Multiplication Puzzles ✦ Activity 53

Correction Key

Swans

1. Allow your students to correct their own work.
2. Make a transparency of this puzzle and instruct your students to place the transparency over their completed puzzle for a quick and easy check.

Multiplication Puzzles ♦ Activity 54

108

© Golden Educational Center

Swans

Name _____

Date _____

1. Complete the problems within the parentheses first. Then complete the subtraction problem using your answer. (You can do the problems on another piece of paper.)
2. Using your final answer and the color key, color your puzzle correctly.

48 Total Problems

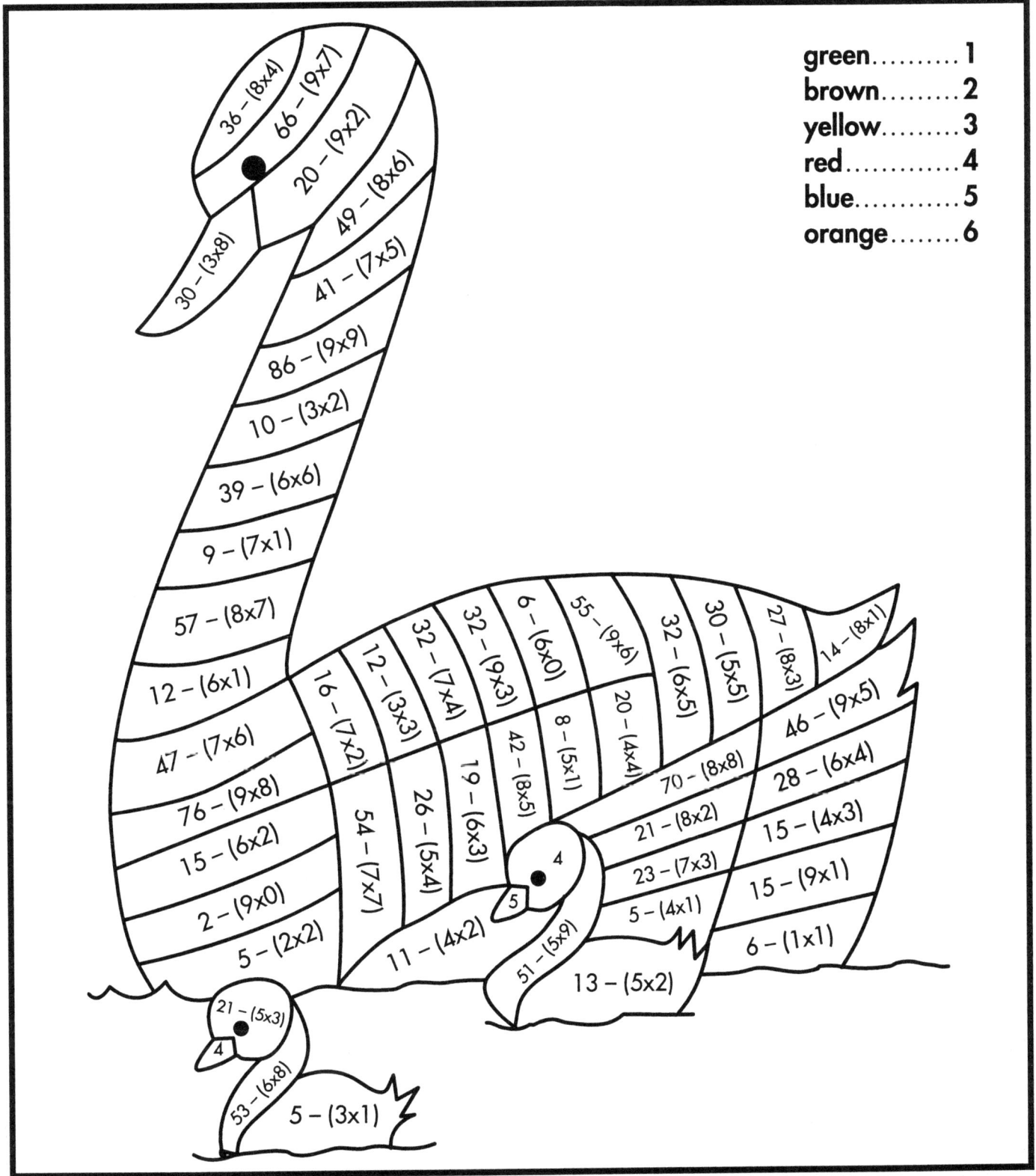

green.........1
brown.........2
yellow.........3
red............4
blue...........5
orange........6

Multiplication Puzzles ✦ Activity 54

Correction Key

Teddy Bear

1. Allow your students to correct their own work.
2. Make a transparency of this puzzle and instruct your students to place the transparency over their completed puzzle for a quick and easy check.

Multiplication Puzzles ✦ Activity 55

Teddy Bear

Name _____

Date _____

1. Complete the problems within the parentheses first. Then complete the addition or subtraction problem using your answer. (You can do the problems on another piece of paper.)
2. Using your final answer and the color key, color your puzzle correctly.

67 Total Problems

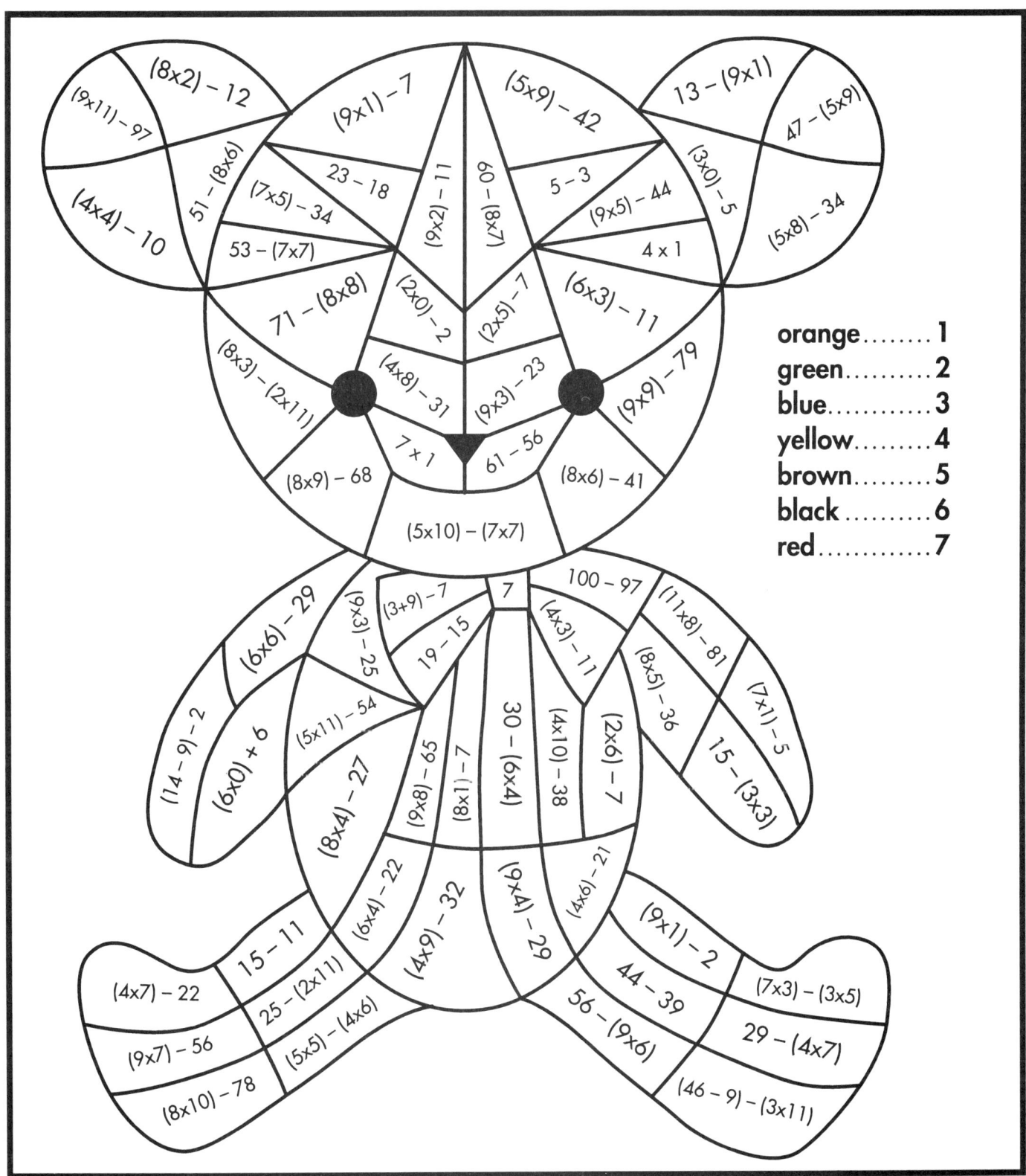

orange........1
green.........2
blue..........3
yellow........4
brown.........5
black.........6
red...........7

Multiplication Puzzles ✦ Activity 55

Correction Key

Tepee

1. Allow your students to correct their own work.
2. Make a transparency of this puzzle and instruct your students to place the transparency over their completed puzzle for a quick and easy check.

Multiplication Puzzles ✦ Activity 56

© Golden Educational Center

Tepee

Name _____

Date _____

1. Complete the problems within the parentheses first. Then complete the subtraction problem using your answer. (You can do the problems on another piece of paper.)
2. Using your final answer and the color key, color your puzzle correctly.

67 Total Problems

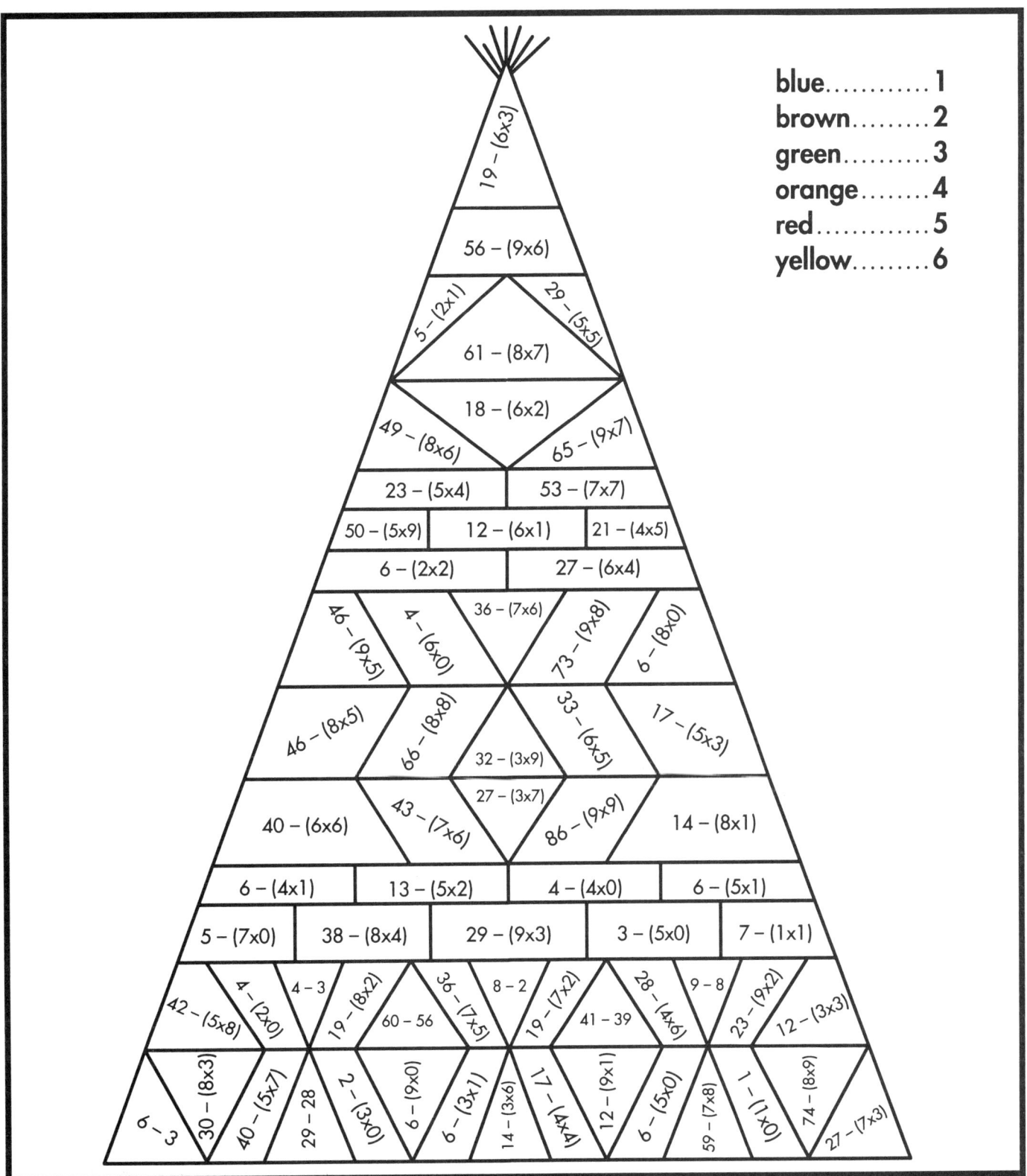

blue............1
brown.........2
green..........3
orange........4
red.............5
yellow.........6

Multiplication Puzzles ♦ Activity 56

Correction Key

Tiger Cub

1. Allow your students to correct their own work.
2. Make a transparency of this puzzle and instruct your students to place the transparency over their completed puzzle for a quick and easy check.

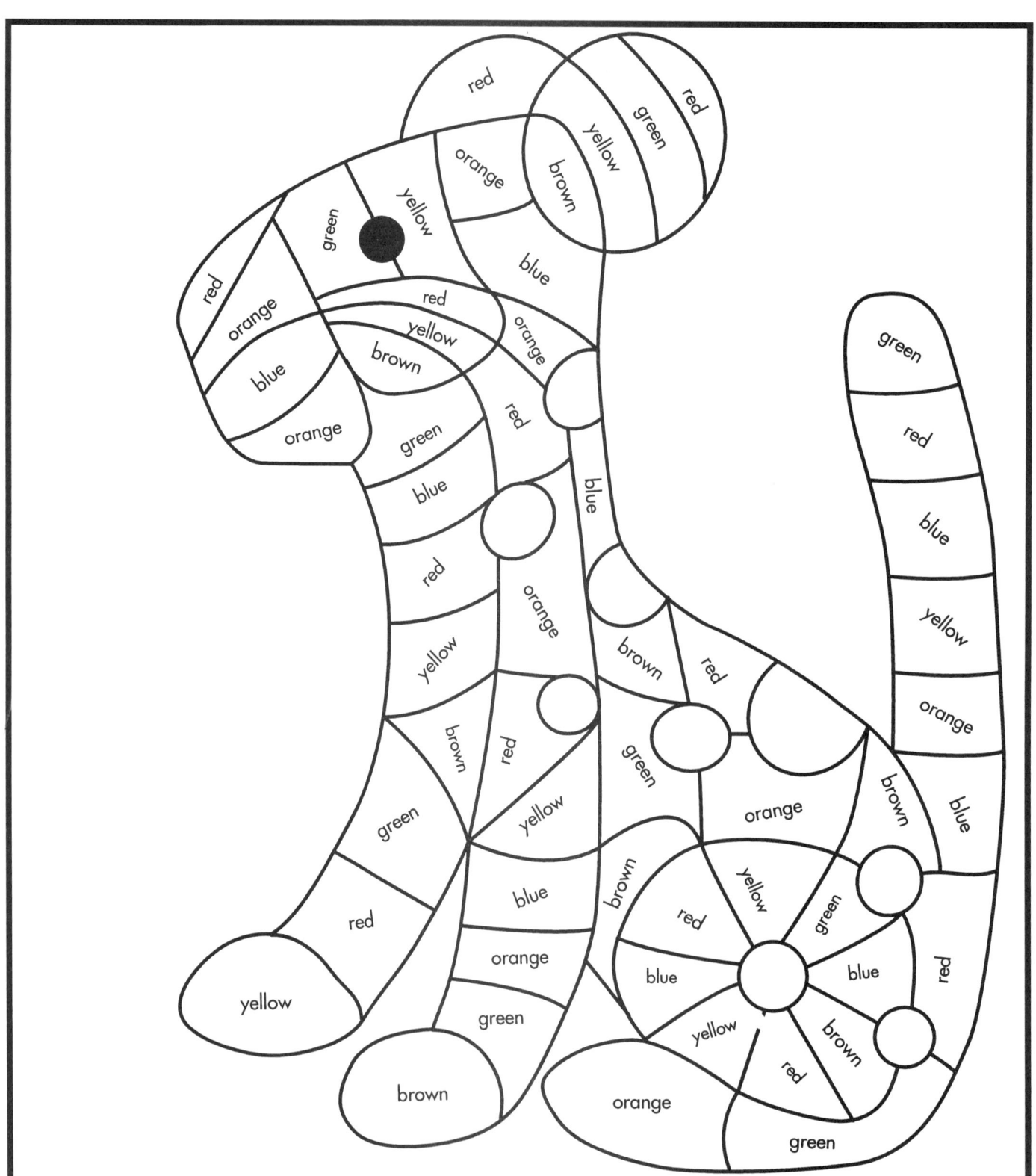

Multiplication Puzzles ✦ Activity 57

Tiger Cub

Name _____

Date _____

1. Complete the problems within the parentheses first. Then complete the subtraction problem using your answer. (You can do the problems on another piece of paper.)

2. Using your final answer and the color key, color your puzzle correctly.

56 Total Problems

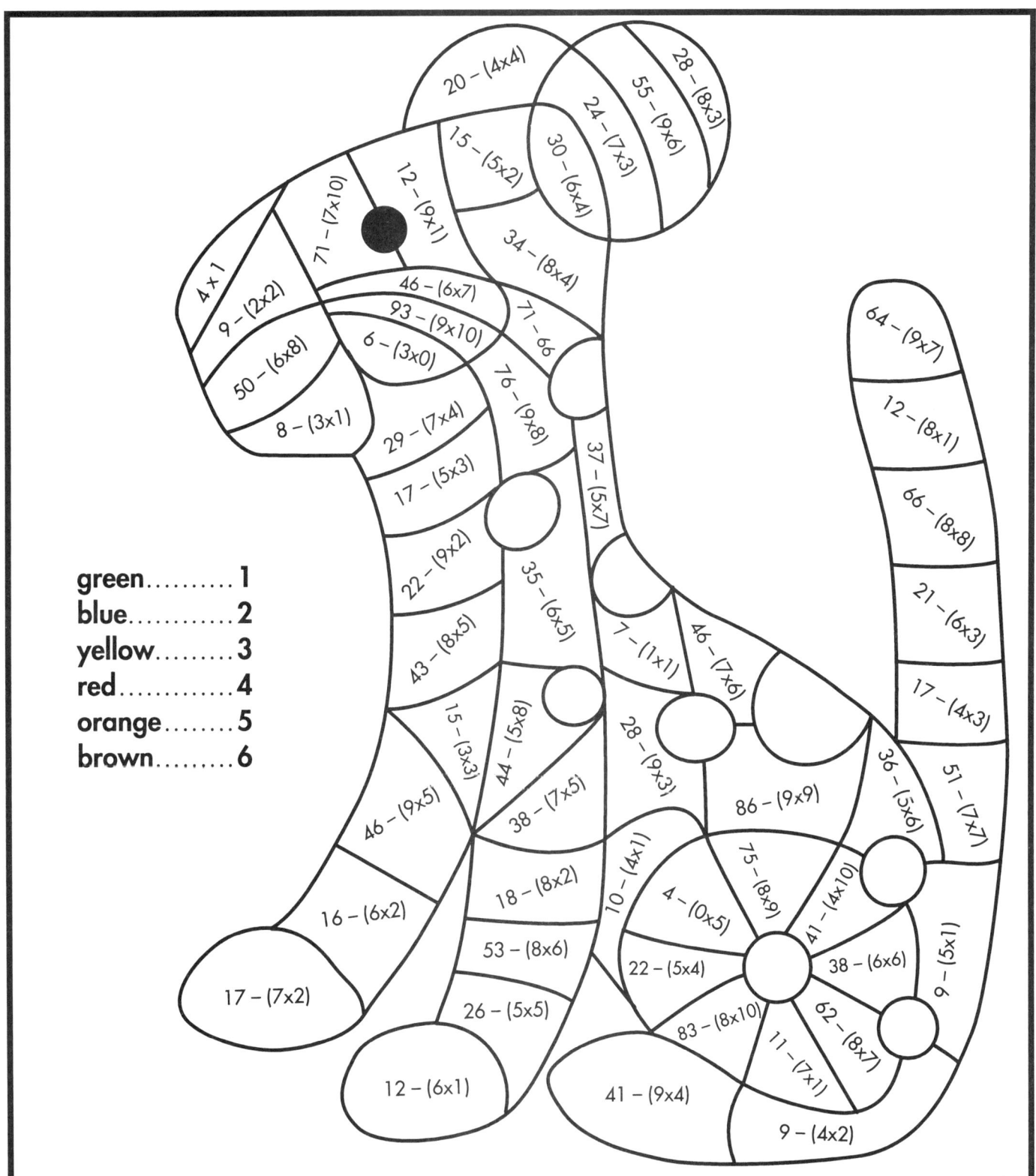

green.........1
blue...........2
yellow........3
red............4
orange.......5
brown........6

Multiplication Puzzles ✦ Activity 57

Correction Key

Tree

1. Allow your students to correct their own work.
2. Make a transparency of this puzzle and instruct your students to place the transparency over their completed puzzle for a quick and easy check.

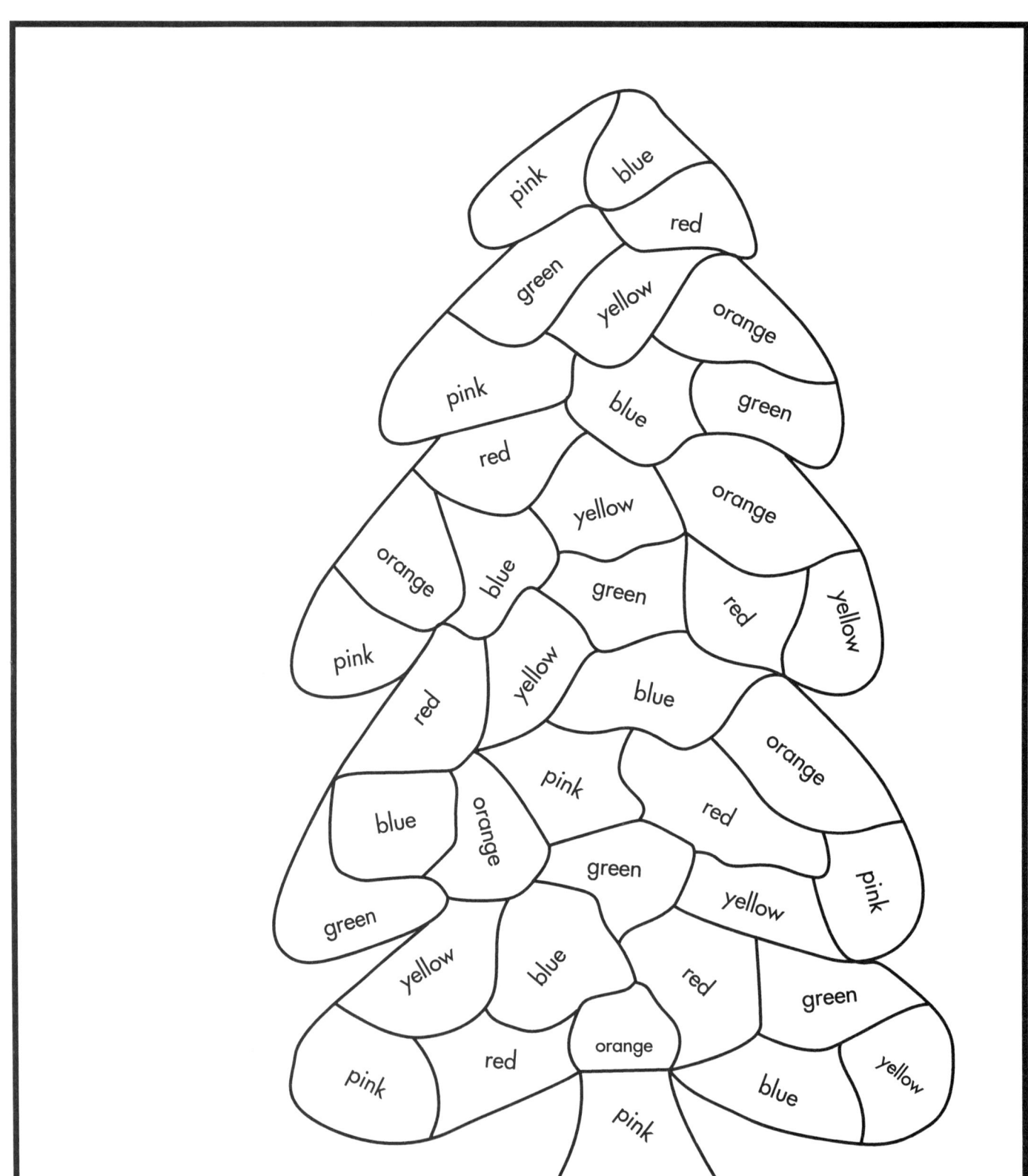

Multiplication Puzzles ✦ Activity 58

Tree

Name _____

Date _____

1. Complete the problems within the parentheses first. Then complete the subtraction problem using your answer. (You can do the problems on another piece of paper.)

2. Using your final answer and the color key, color your puzzle correctly.

40 Total Problems

pink..........1
blue..........2
red...........3
green........4
yellow.......5
orange.......6

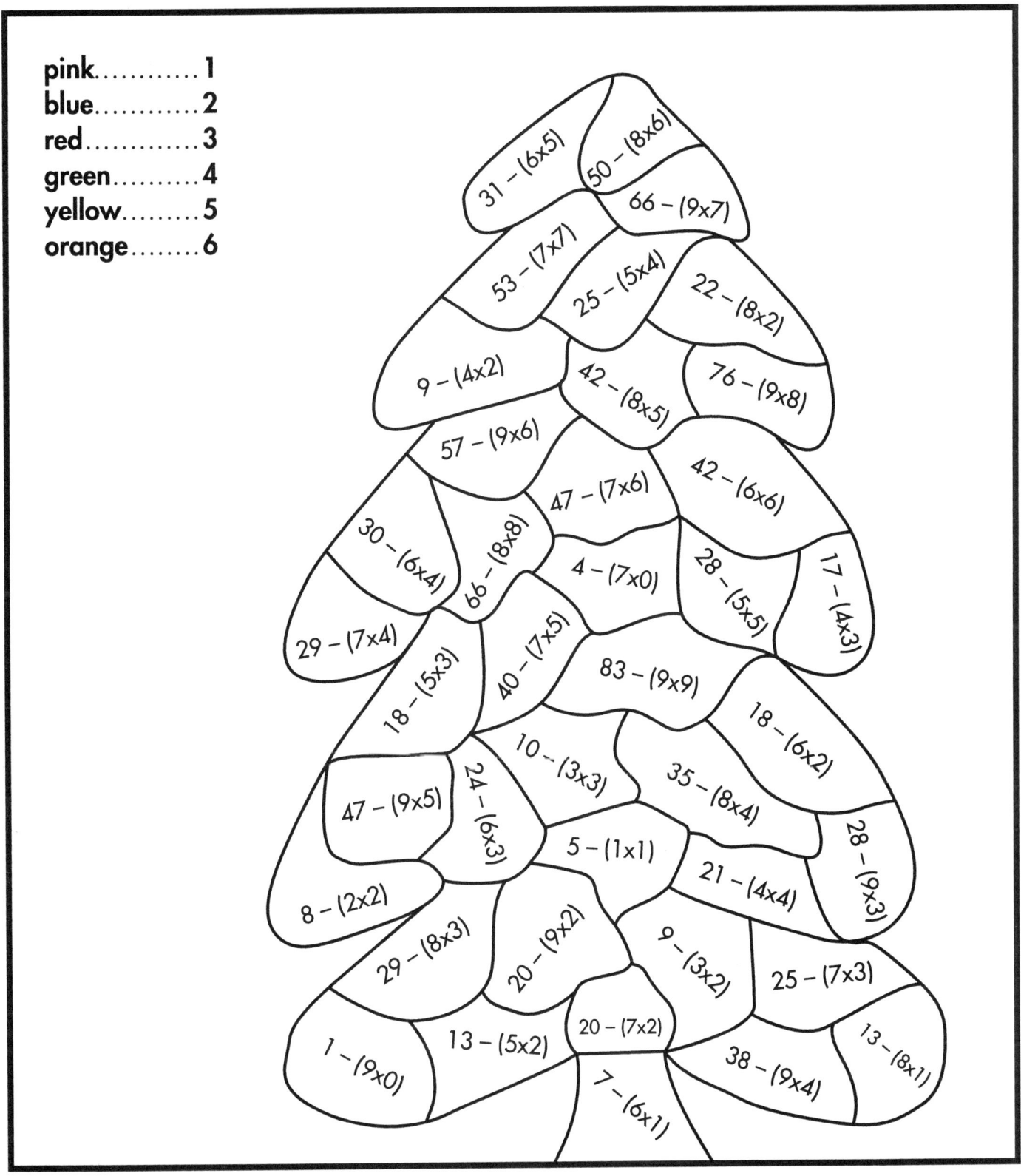

Multiplication Puzzles ✦ Activity 58

Correction Key

United States of America

1. Allow your students to correct their own work.
2. Make a transparency of this puzzle and instruct your students to place the transparency over their completed puzzle for a quick and easy check.

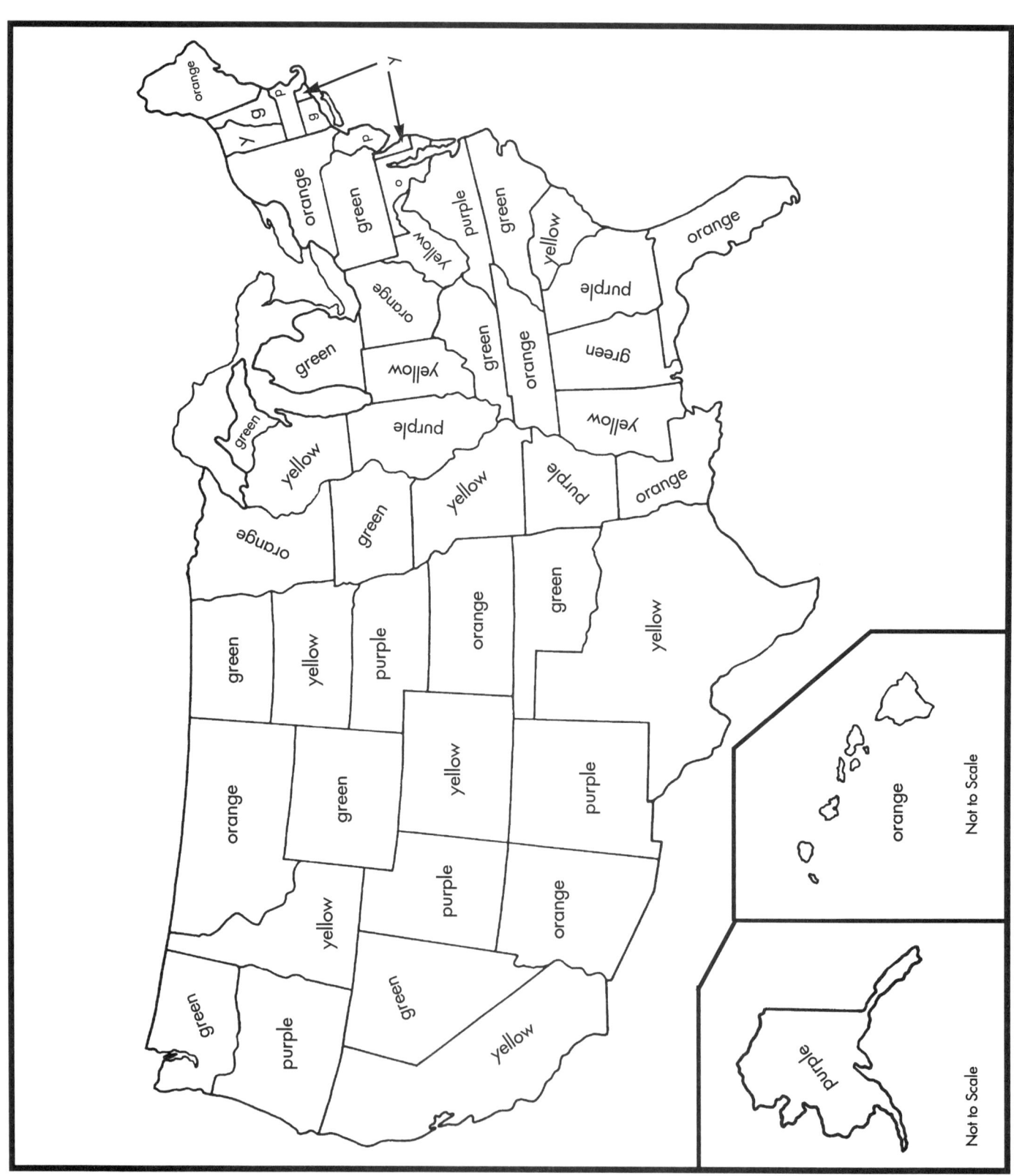

Multiplication Puzzles ✦ Activity 59

United States of America

Name _____

Date _____

1. Complete the problems within the parentheses first. Then complete the addition or subtraction problem using your answer. (You can do the problems on another piece of paper.)
2. Using your final answer and the color key, color your puzzle correctly.

43 Total Problems

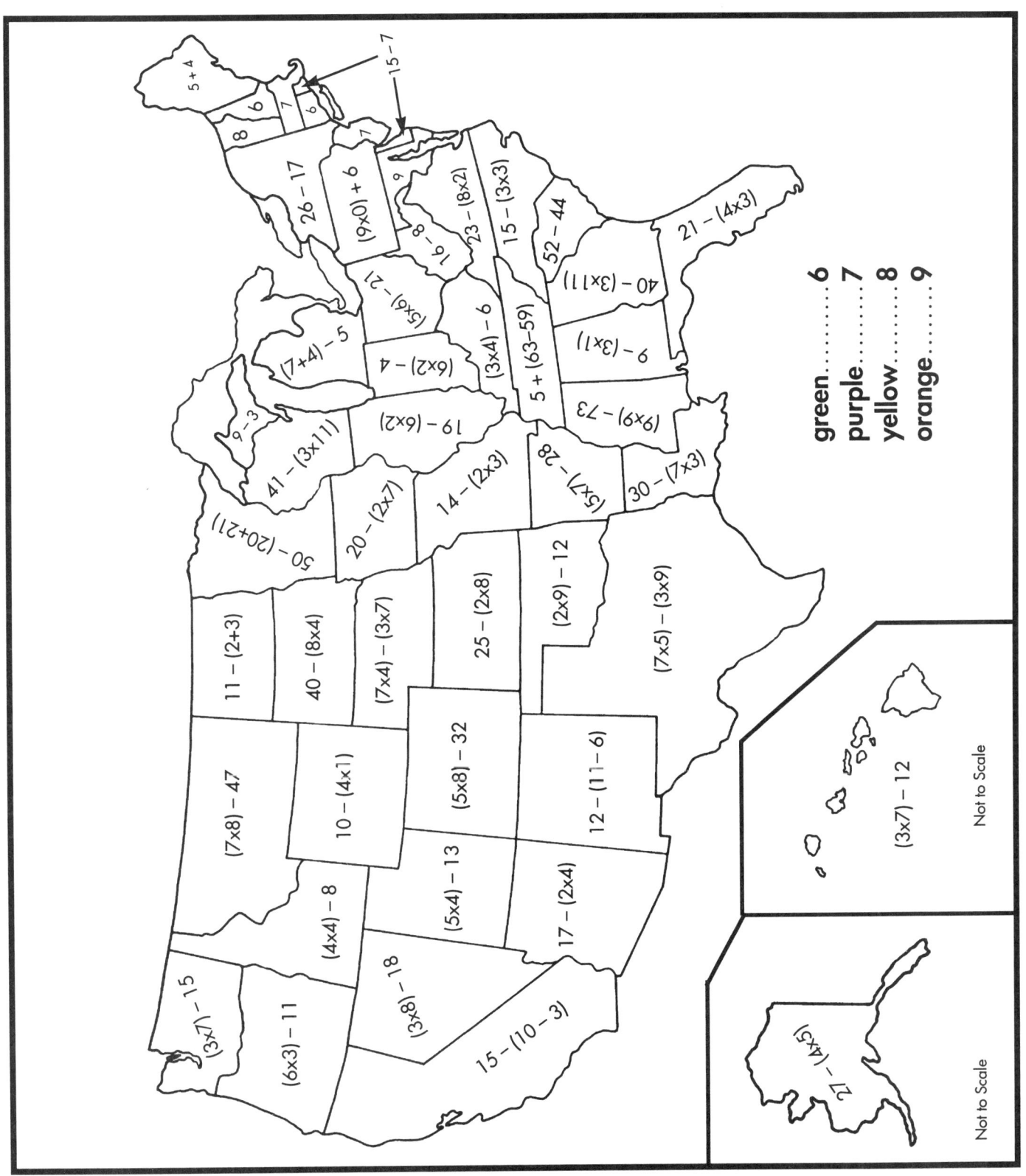

Multiplication Puzzles ✦ Activity 59

Correction Key

Valentine

1. Allow your students to correct their own work.
2. Make a transparency of this puzzle and instruct your students to place the transparency over their completed puzzle for a quick and easy check.

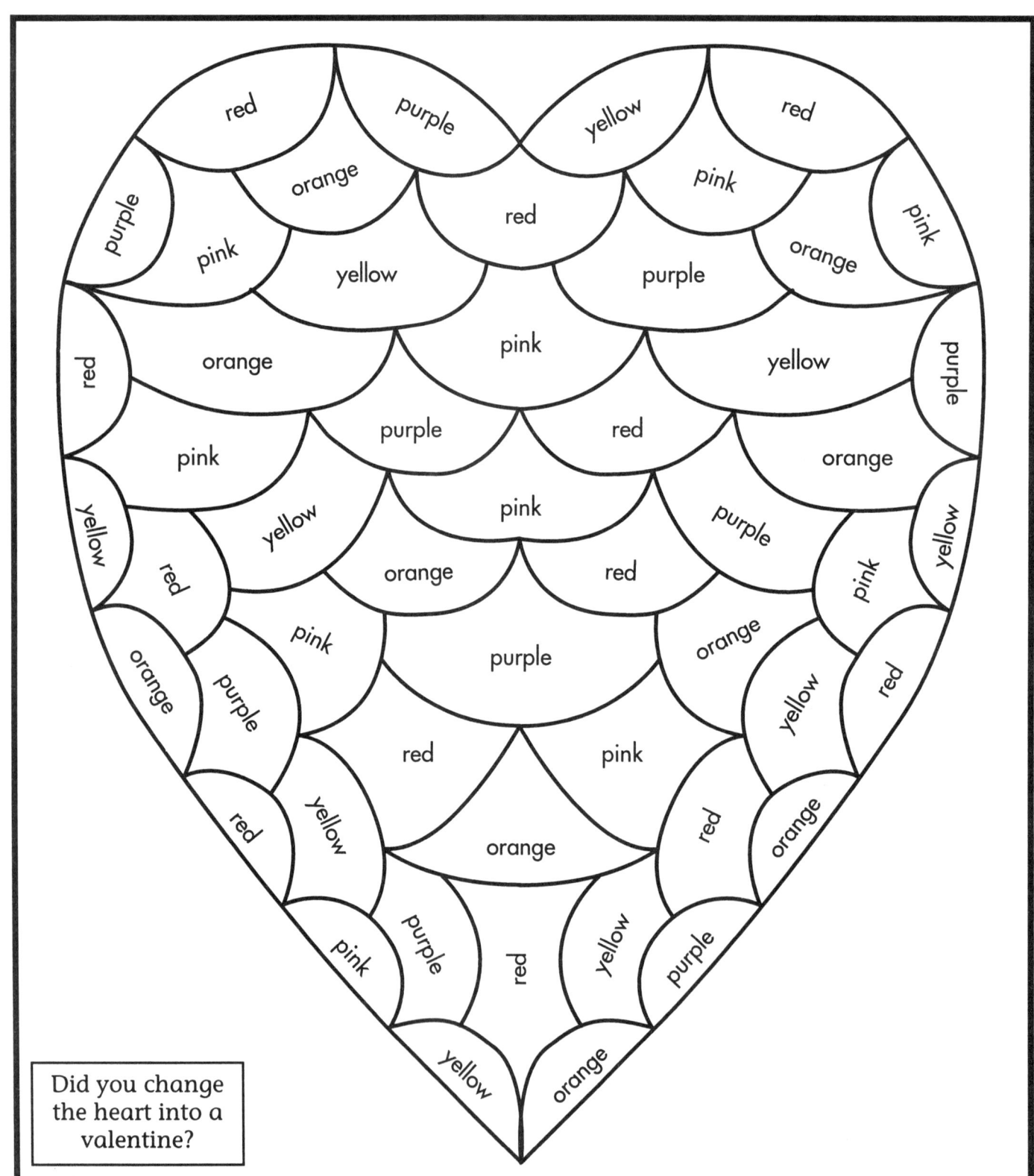

Did you change the heart into a valentine?

Multiplication Puzzles ✦ Activity 60

Valentine

Name _____

Date _____

1. Complete the problems within the parentheses first. Then complete the addition or subtraction problem using your answer. (You can do the problems on another piece of paper.)

2. Using your final answer and the color key, color your puzzle correctly.

52 Total Problems

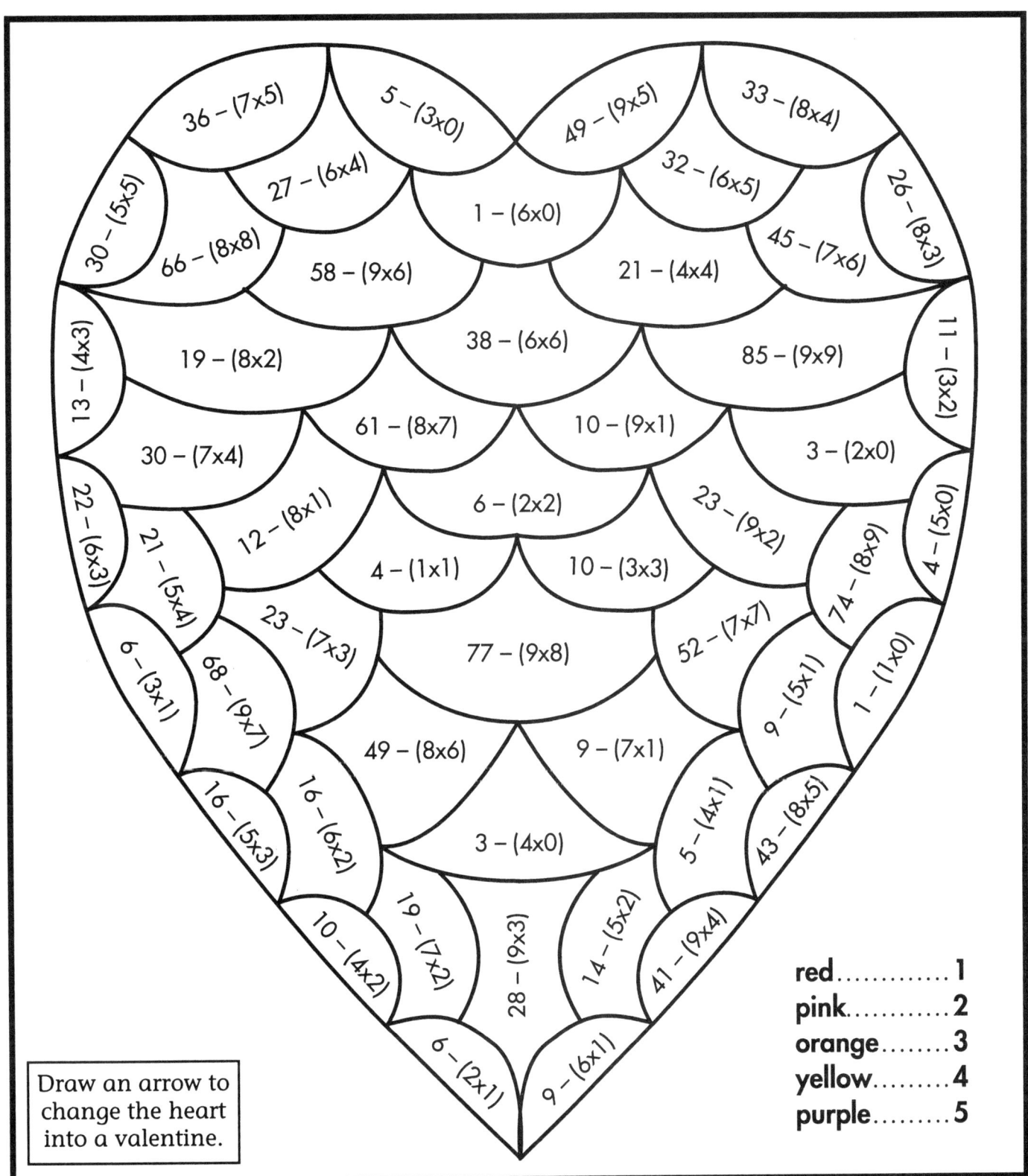

red............1
pink...........2
orange........3
yellow........4
purple........5

Draw an arrow to change the heart into a valentine.

© Golden Educational Center   121   Multiplication Puzzles ✦ Activity 60

Notes & Doodles

Multiplication Puzzles ◆ **Notes & Doodles**

Mutiplication Grid

As the teacher, you can decide whether or not your students should be allowed to use the multiplication grid below. We wanted to include it just in case you have students who need it.

0	1	2	3	4	5	6	7	8	9
1	1	2	3	4	5	6	7	8	9
2	2	4	6	8	10	12	14	16	18
3	3	6	9	12	15	18	21	24	27
4	4	8	12	16	20	24	28	32	36
5	5	10	15	20	25	30	35	40	45
6	6	12	18	24	30	36	42	48	54
7	7	14	21	28	35	42	49	56	63
8	8	16	24	32	40	48	56	64	72
9	9	18	27	36	45	54	63	72	81

More Notes & Doodles

GOLDEN EDUCATIONAL CENTER
G.E.C. PUBLICATIONS
"LEADING THE WAY IN CREATIVE EDUCATIONAL MATERIALS"™

857 LAKE BLVD. REDDING, CA 96003

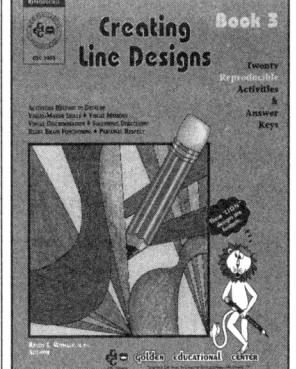

CREATING LINE DESIGNS

These are a well-liked series of workbooks that incorporate using a pencil and ruler to connect points on a page. In the *CLD* books, a completed pattern is shown as an example for the child to duplicate. Each book is progressively more difficult in design, with 20 designs per book. Students are motivated to master more difficult designs and consequently to develop their visual perception and memory. We've also called them "pre-drafting skill workbooks," as they are designed to get students to master ruler and pencil skills.

BOOK 1 #1001	K-1st
BOOK 2 #1002	2nd-3rd
BOOK 3 #1003	3rd-6th
BOOK 4 #1004	4th-7th

❖ ❖ ❖

DESIGNS IN MATH

Students create geometric designs by connecting the dots with a ruler between a math problem and the correct answer. These activities reinforce basic math fact memorization, following directions, fine muscle control, visual-motor skills, and principles of design. Each book contains 20 reproducible activities. There is a completed design that can be used for a correction key or made into a transparency for classroom instruction.

ADDITION #1006	3rd-5th
SUBTRACTION #1007	3rd-5th
MULTIPLICATION #1008	3rd-6th
DIVISION #1009	4th-7th
FRACTIONS #1010	5th-9th
FRAC.-DEC. EQUIVALENTS #1011	5th-9th

❖ ❖ ❖

READ•N•DRAW
FOLLOWING DIRECTIONS

These great workbooks teach children to follow directions through exciting measuring/drawing activities. Students increase reading comprehension by reading sequential directions, plotting points and completing the Twenty activities in each book make learning fun. *Teacher instructions and keys are included.*

| BOOK 1 #1021 | 3rd-5th |
| BOOK 2 #1022 | 4th-8th |

❖ ❖ ❖

BEGINNING MATH ART

Similar to *Designs in Math,* these books show students the fundamentals of design while they practice solving their math problems and then connect the correct dots with a straight edge. There is a maximum of 12 problems per design. Young students will enjoy completing the 20 activity pages in each of these books. *Teacher instructions and answer keys are included.*

ADD & SUBTRACT 0-10 #1013	K-1st
ADD & SUBTRACT 11-20 #1014	1st-2nd
MULT. & DIVIDE 0-12 #1015	2nd-3rd

❖ ❖ ❖

USING A VISUAL GRID FOR SOLVING MATH WORD PROBLEMS

Paring down to the most essential information, young students can now see and even understand the most essential elements/terms of a word problem. There are 30 lessons in each of these books. *Teacher instructions and answer keys are included.*

ADD & SUBTRACT 0-99 #2221
(no borrowing/carrying)   K-2nd

ADD & SUBTRACT TO 999 #2222
(with borrowing/carrying)   1st-3rd

MULT. & DIVISION #2223   3rd-5th

MEASUREMENTS, TIME & MONEY
Mostly Mult. & Div.)   4th-7th

❖ ❖ ❖

MULTIPLICATION PRACTICE PUZZLES

The activities in this book have students complete a multiplication problem and then a subtraction (or addition) problem using the answer to the multiplication problem. After determining the final answer, they color the shape containing the problems the correct color, which is determined by the color key on the activity page. Each picture has 30 to 50 problems per page. The book has 60 different puzzles (pictures) and answer keys.

GRADES 3RD-7TH #1025

ECOLOGY MATH

Students are presented with simplified ecological information. They then write their ideas on how they can help the ecological situation. The lessons continue with mathematical problems referring to the information first given. There are 17 math lessons, as well as 17 discussion and information sheets. *Teacher instructions and answer keys are included.*

GRADES 5TH-8TH #1105

857 LAKE BLVD. REDDING, CA 96003

COUNTRY STUDY BOOKS

Each individual country within each respective continent has new words to look up in a dictionary, a large outline map, one page of current facts (population, area, capital, largest city, flag w/description, additional interesting information and more), one or two pages of history through independence and a page of review questions at the end of each section. Bonus (research) activities are also at the end of each section. *96 to 112 pages; Teacher instructions and keys are provided.*

NORTH AMERICA #1965 .... 4th-8th
SOUTH AMERICA #1975 .... 4th-8th
FAR EAST #1935 .... 4th-8th
MIDDLE EAST #1936 .... 4th-8th
CANADA #1985 .... 4th-8th

❖ ❖ ❖

CALIFORNIA ❖ WASHINGTON ❖ TEXAS

California Early History:
This is a simplified, yet complete, resource book detailing the history of California through statehood. This tremendous resource includes sections on the contributions made by Native Americans, Asians, Africans and Mexicans to the growth and development of the state. Review questions and bonus (research) activities follow each section. *94 pages; Teacher instructions and answer keys are provided.*

California Geography:
Students are given a world overview with a specific review of California's climate and its physical, economic and political features. Lessons are reinforced with maps, exercises, review questions and bonus (research) activities. *63 pages; Teacher instructions and answer keys are provided.*

Washington State Geography:
Identical to California, but with Washington's information. *63 pages; Teacher instructions and answer keys are provided.*

Texas Geography:
Identical to California, but with Texas information. *63 pages; Teacher instructions and answer keys are provided.*

California Missions & California Missions Poster:
Coming Summer 1998

CALIFORNIA EARLY HISTORY #2911 .... 4th-8th
CALIFORNIA GEOGRAPHY #2912 .... 4th-8th
CALIFORNIA MISSIONS #2913 .... 4th-8th
CALIFORNIA MISSIONS POSTER #2914
WASHINGTON GEOGRAPHY #2106 .... 4th-8th
TEXAS GEOGRAPHY #2105 .... 4th-8th

LEARNING THE CONTINENTS

Students use maps to identify, memorize and locate the countries, waterways and points of interest on each continent (Australia and Antarctica are not divided into countries). Each continent has 16 pages of map activities and a word search puzzle at the end for spelling review of the countries. *96 pages; Teacher instructions and answer keys are provided.*

#1906 .... 4th-8th

❖ ❖ ❖

CONTINENT MAPS & STUDIES

This book contains outline, waterway and political boundary maps and an individual fact sheet for each of the continents. (Antarctica only has an outline map and fact sheet.) There are questions, research activities and a glossary that can be used with each continent. *64 pages; Teacher instructions and keys.*

#1905 .... 4th-8th

❖ ❖ ❖

U.S. STUDIES

U.S. Outline Maps:
This book has an individual fact sheet and outline map for each state, Washington, D.C. and the entire United States. U.S. waterways and state boundary maps are also included. There are also individual question and research pages that can be used with each of the states. *112 pages.*

U.S. Geography:
Each section of this book begins with new vocabulary words to look up and define. The sections cover a world overview and physical, economic, political and climatic features of the country. Map activities, review questions and bonus (research) activities are included in each section. There is also a review section for the entire book with crossword puzzles, a word search and other activities. *80 pages; Teacher instructions and keys.*

Complete Book of U.S. State Studies:
This book includes the same information as in the *U.S. Outline Maps* book listed above. It also includes historical information about Native Americans living in the region before any settlers came, as well as the early settlers up to the time of statehood. Each state also has interesting trivia facts and a word search puzzle. *219 pages; Teacher instructions and keys are included.*

U.S. OUTLINE MAPS #1992 .... 4th-8th
U.S. GEOGRAPHY #1993 .... 4th-8th
COMPLETE U.S. STATE STUDIES #1995 .... 4th-8th